AKG

OP

DABIESHAN

Scandinavian Institute of Asian Studies
Monograph Series

China:

No. 16	Trygve Lötveit: Chinese Communism, 1931-1934 Sources for the understanding of contemporary China
No. 21	Robert Tung: Proscribed Chinese Writing A textbook for students of the modern Chinese language
No. 26	Wang Eang Cheong: Mandarins and Merchants A China agency of the early nineteenth century
No. 38	Hans J. Hinrup: An Index to *Chinese Literature* 1951-1976 Alphabetical and subject indexes to this Chinese journal
No. 42	N.H. Leon: Character Indexes of Modern Chinese Reflecting recent developments of the Chinese language
No. 43	Göran Aijmer: Economic Man in Sha Tin Chinese immigrant gardeners in a Hong Kong valley
No. 45	Christoph Harbsmeier: Aspects of Classical Chinese Syntax The basic problems of classical Chinese grammar
No. 50	S.A.M. Adshead: Province and Politics in late Imperial China Viceregal government in Szechwan, 1898-1911

SCANDINAVIAN INSTITUTE OF ASIAN STUDIES
MONOGRAPH SERIES NO. 52

Dabieshan

Traditional Chinese iron-production techniques practised in southern Henan in the twentieth century

Donald B. Wagner

Curzon Press

Scandinavian Institute of Asian Studies
Kejsergade 2, DK-1155 Copenhagen K

First published 1985

Curzon Press Ltd : London and Malmö

ISBN 0 7007 0177 X
ISSN 0069 1712

Printed by Narayana Press, Gylling, DK-8300 Odder, Denmark

Dedicated to my father
William J. Wagner

CONTENTS

PREFACE

Many iron-production techniques of surprising antiquity survived in China well into modern times, and have been described in detail by competent modern engineers. It is my contention that a careful study of traditional Chinese techniques as they have been practised in the twentieth century will provide a basis for the interpretation of archaeological remains and written records, and thus contribute to the study of the ancient Chinese iron industry: both its technical development and its economic history.

Considering the enormous size and geographical variation of China, and the long history of its iron industry, it should not be surprising that a wide variety of iron-production techniques have been in use. Each region has had its own local techniques, adapted to its particular ecological, economic, and social conditions. Each must be considered separately.

The present book, intended to be the first of several studies, focuses on the Dabieshan mountains, a rugged, isolated region comprising parts of the provinces of Henan, Hubei, and Anhui. In 1917 it was estimated that there were at least 100 ironworks in operation in the Henan part of the region, with an output of about 14000 tons per year. The raw materials were ironsand and charcoal. Pig-iron was produced in a small blast furnace, only two metres high, and converted to wrought-iron in a very small hole-in-the-ground refining hearth.

The Dabieshan region seems never to have had much importance in Chinese history, and the choice of its iron industry as the place to begin these studies has largely been determined by the availability of sources. A brochure published in connection with the Great Leap Forward in 1958, translated in chapter 2 below, considers most aspects of this industry, and it has been possible to supplement this description with various others, from 1917, 1932, and 1958. Thus it has been possible to study the local iron industry in all its phases, from ironsand sluicing and charcoal production through the blast furnace and the refining furnace, and see how these processes fit together as a whole.[1] The fact that the techniques used are comparatively simple, and show only a comparatively narrow range of variation within the region, has also contributed to the choice.

The sources are available for similar studies of local iron industries in many parts of China. Perhaps the most promising is that of the province of Sichuan: the techniques are considerably more complex (including some use of coke as the fuel, and of water-power to drive the blast), and a great many descriptions are available. Furthermore Sichuan's history and economic geography have been studied by many Chinese and foreign scholars, so it should be possible to place the iron industry in a broader context. In the present study it has been impracticable to include this broader context because the Dabieshan region has not been much studied. The sources for such a study are available in local gazeteers etc., but the rigorous use of this material would require a major research effort.

Chapter 1 provides some introductory orientations on the Dabieshan region and on the sources for the study of its iron-production technology. Chapter 2, the translation mentioned above, gives a general description of the iron industry of the region. Following chapters take up the individual techniques of the industry in turn: ironsand-sluicing, charcoal production, smelting, and refining. In these chapters the primary concern is to provide additional technical detail from other sources; in addition I have attempted to place each technique in a wider context, considering briefly some of the variations to be found in other parts of China, and some historical aspects.

This book is primarily a technical study, and large parts of it will make difficult reading for those who have no knowledge at all of extractive metallurgy. Nevertheless I hope that this detailed description of a particular industry in a particular place will provide useful material for economic historians of modern China. I hope too that parts of it will be read by historians of contemporary China, and that it will provide some nuances for discussions of the 'Backyard Furnace Campaign' of 1958. There is no doubt that that campaign was a political and economic mistake, but I believe that exactly what happened, and exactly why it was a mistake, are more complex questions than has usually been recognized. Aware of my own limitations, however, I must leave this interesting matter to specialists.[2]

Another group which this book may reach is the amateur experimenters: with the information provided here it should be possible, with a fair amount of trial and error, to build a backyard ironworks. I hope experimenters will take all necessary precautions, and that no serious accidents occur. And I certainly hope that anyone who tries will let me know how things turn out.

All Chinese names and terms are transcribed according to the *Pinyin* system. The table of Chinese characters provides equivalents in the Wade-Giles system. All place-names mentioned are indicated on the map, figure 14.

This book is part of a study financed by the Danish Research Council for the Humanities, which also financed its publication. It was written while I was a Visiting Scholar at Stanford University and the University of California, Berkeley, where I discovered the joys of working in two truly great libraries. It is a pleasure to acknowledge my indebtedness to Dr. Vagn Fabritius Buchwald, lecturer in metallurgy at the Technical University of Denmark. He not only taught me what I know of metallurgy, but he has also continually supported my work with advice and encouragement. Thanks and an apology are also due to Prof. R.F. Tylecote, editor of the *Journal of the Historical Metallurgy Society*. I had originally written this book as an article in two parts for that journal, and Prof. Tylecote did a great deal of work and provided much valuable advice before I rudely withdrew it to turn it into a book. Neither is, of course, responsible for errors and misunderstandings that remain: especially since neither has seen the final revised version of the book. Annie Winther tends to react with deep-throated feminist laughter to the male academic's traditional acknowledgement of the patient support etc. of his long-suffering etc. wife. Here I shall restrict myself to saying: I love you.

CHAPTER 1

SOURCES ON THE IRON INDUSTRY OF SOUTHERN HENAN

The Funiushan, Tongbaishan, and Dabieshan mountain ranges form the border between the provinces of Henan and Hubei, and extend eastward into Anhui. They form a geographical region rather different from the regions to the north and south. It is characterized by old low mountains, seldom exceeding 1500 m, with fertile valleys extending inward from the plains on either side. The region is rugged and isolated; this is no doubt one reason why one of the first Communist base areas, the E-Yu-Wan Soviet, was established here in 1929.[1]

For the reader interested in the traditional iron industry of China the place to start is Felix Reinhold Tegengren's monumental *The iron ores and iron industry of China* (1924). A chapter titled 'The primitive native industry' (pp. 297-365) reviews the methods of the traditional iron industry province by province, using descriptions by earlier writers as well as Tegengren's own observations.

Tegengren's description of the iron industry of southern Henan is based on a report by another Swedish geologist, Erik Torsten Nyström, who visited the region about 1917 for the Geological Survey of China. Tegengren refers (pp. 179, 334) to 'a report by E.T. Nyström 1917', but gives no further details. I have expended considerable effort in an unsuccessful search for this report: perhaps it was never published.[2]

Another study of the iron industry of southern Henan was by Guo Yujing some time before 1932 for the Henan Provincial Geological Survey. Here again I have not been able to locate the original report, but it is quoted in a later publication.[3] Guo Yujing's visit to the region is described as follows:

> Mr. Guo Yujing was sent to investigate the ironsand of Xinyang. At the time the bandits had not yet been pacified; carrying his own provisions he entered the mountains and carried out observations for more than ten days. He discussed the local iron-smelting methods in detail, and looked to improving them. In the end, because his proposals were inadequately implemented, the results were not great. (Zhang Youxian & Guo Yujing 1932, p. 223)

The mention of 'bandits' here no doubt refers in fact to the E-Yu-Wan Soviet, but there were also real bandits in the region at this time, a group called the Red Spears (Wales 1952, p. 153). The character of the region, with rugged mountains in close proximity to settled agriculture, may have made it a haven for bandits and revolutionaries at all times, but I have not been able to find out more on this subject.

After 1949, with the coming of peace, political stability, and national reconstruction after decades of war, the demand for iron and steel increased enormously, and China's modern sector could not expand rapidly enough to meet this demand. In many parts of China, especially in the lower Changjiang (Yangtze) Valley, small ironworks were established by local authorities to help meet local requirements (Zhang Zongling 1957). These were in most cases based on the techniques of the traditional industry of these localities, which in spite of a deep decline had not been totally extinguished.

Compared with modern ironworks these old-fashioned ironworks had both advantages and disadvantages. They required a much smaller initial investment and were built using local materials. The necessary skilled workers were already available, and their skills could be acquired on the job fairly quickly by new workers. Advantage could be taken of small sources of raw materials which could not be exploited economically by modern large-scale methods. The problems of social dislocation associated with large-scale industry could be avoided. The burden on China's still-inadequate transportation facilities could be alleviated. For local authorities it must have been a great advantage to have a local supply of iron, for reliance on outside supplies must have involved difficult political negotiations.

The disadvantages lay in both the quality of the product and its cost. The small wrought-iron bars produced in the ironworks of the Dabieshan region (see figure 13 and the chemical analyses given in chapter 2, section 2.3.3, and chapter 6) were undoubtedly an excellent material for blacksmiths producing small agricultural implements and the like; but they could hardly be used in a modern factory. The description translated in chapter 2 provides data on the labour required for all phases of iron production by the Dabieshan methods; from these it is possible to estimate that the labour required to produce one ton of wrought-iron bars (in ironsand-sluicing, charcoal production, smelting, and refining) was of the order of 2-4 worker-months. Thus this iron was much more expensive than that produced by modern methods.[4]

On balance it seems that the temporary expansion of the traditional sector was a sensible solution to some pressing problems which confronted China in the 1950s. But then came the Great Leap Forward of 1958-60 and the related campaign which Western journalists have called the 'Backyard Furnace Campaign'. It was announced in September 1958, and involved the building of hundreds of thousands of small-scale ironworks all over China. It was a fiasco. Today the official Chinese evaluation of the Great Leap Forward is very harsh:

> Its shortcoming was that it overlooked objective economic laws. Both before and after the plenum,[5] all comrades . . . displayed high enthusiasm and initiative for socialism and achieved certain results in production and construction. However, 'Left'

errors, characterized by excessive targets, the issuing of arbitrary directions, boastfulness and the stirring up of a 'communist wind', spread unchecked throughout the country. This was due to our lack of experience in socialist construction and inadequate understanding of the laws of economic development and of the basic economic conditions of China. ... Comrade Mao Zedong and many leading comrades ... were impatient for quick results and overestimated the role of man's subjective will and efforts.

(*Resolution* 1981, p. 28)

My own scattered reading of the sources for the 'Backyard Furnace Campaign' generally confirms this evaluation: enthusiasm for an idea was in command, and the voices of realists, though audible, were not heard. On the other hand a great deal of important technical work was done, and this seems to have led to the development of small modern ironworks throughout China which are much more efficient than the traditional works and better adapted to present conditions than large-scale works based on Western models.[6]

No matter how the political and economic aspects of the 'Backyard Furnace Campaign' are to be evaluated, it left behind a marvellous body of documentation of traditional Chinese iron-production techniques. They are described in considerable detail in a great many technical publications, and these have been the most important sources in the present study of the Dabieshan iron industry. The technical studies were intended for practical use, and thus tend to be much better sources than such descriptions as those of Nyström and Guo Yujing, which treat the traditional methods more as curiosities than as immediate practical concerns.

Nyström estimated that there were about 100 ironworks in southern Henan: 15 in Xinyang, 10 in Guangshan, and 75 in Shangcheng; they produced about 14 400 tons of iron per year (Tegengren 1924, p. 334). Guo Yujing estimated the annual production as 200 tons in Xinyang and 4600 tons in Shangcheng; for Guangshan no estimate was possible. From these figures one might conclude that production had dropped considerably between 1917 and 1932; especially in Xinyang, which is on the Beijing-Hankou Railway (completed in 1906) and therefore would be the first place in the region to be affected by competition with cheap foreign iron.

Nevertheless such figures should be treated with great caution. We do not know how they were arrived at, and we can imagine the difficulties of an observer travelling on foot or horseback for a few days in a mountainous region larger than Denmark. The estimates of both Nyström and Guo Yujing should be taken as minimum figures, representing only those ironworks which the observer happened to learn about.

The descriptions of the ironworks themselves in 1917, 1932, and 1958 should be seen in the same light. The variations reported probably do not represent developments over time; more likely they are local variations which the three observers happened to see in different ironworks.

For technology in other parts of the region than southern Henan I have found only one description: a 1958 study of a blast furnace in Macheng County, in northern Hubei, discussed below in chapter 5, section 5.5. Here the furnace is very similar to that of southern Henan, but its operation is rather different. It is an interesting question whether this represents a real difference in the techniques used in the areas north and south of the mountains: it is a real possibility, but without more sources the question is best left open.

No doubt the iron industry of this region flourished for many centuries, but I have found no information on its history before the twentieth century. There exists a great deal of early material of possible relevance to this question in local gazeteers, travel accounts, state documents, etc.; a search through a small fraction of this material yielded nothing useful, but a more thorough search would surely yield something.

What is really needed, however, is an archaeological survey of the slag heaps of the region. This would surely give a great deal of information on the early history of the iron industry. At present very little work is being done in China on the industrial archaeology of recent centuries, but we may hope that it is only a matter of time before Chinese archaeologists take up this fascinating study.

For the recent history of the iron industry of the region there must be thousands of living witnesses to be interviewed.

CHAPTER 2

TRADITIONAL IRON-PRODUCTION TECHNIQUES OF SOUTHERN HENAN

During the Great Leap Forward period of 1958-60 a campaign was organized to develop and propagate the traditional small-scale methods of iron and steel production used locally in all parts of China. In this 'Backyard Furnace Campaign' (a term used often in the West but not, as far as I know, ever used in China) it seems to have been recognized that iron and steel produced by traditional methods were more expensive than, and generally inferior to, those produced by modern methods. On the other hand, the establishment of a traditional-style ironworks required a smaller initial investment and a lower level of technical knowledge; furthermore advantage could be taken of numerous sources of rich ore which were too small for economic exploitation by modern methods.

In connection with the campaign over 1000 books and brochures were published.[1] *Roughly one-third of these seem to have been political in content, the rest technical. This material has hitherto been unavailable outside China, but in 1982 Dr. Joseph Needham was able to obtain photocopies of fourteen brochures from the Beijing Library. The following is an almost complete translation of one of these brochures (*Tufa *1958).*

In the translation I have omitted some peripheral remarks, some repetitions, and some information which is quoted from modern engineering textbooks; these omissions amount to about eight of the twenty-six pages of the original. All information on the actual techniques has been translated. Several photographs are referred to in the text, but these were not included in the photocopy available to me. Where possible I have substituted other photographs from a variety of sources, in the hope that these photographs correspond to those in the original. All the notes are mine.

2.1. Introduction

2.1.1. General considerations

The use of iron in China probably goes back 4000 years. Its first use on a large scale, judging from historical sources, was in the Spring and Autumn (8th-5th cent. B.C.) and Warring States (5th-3rd cent. B.C.) periods.[2] Thus the Chinese regional methods of iron production were created by the toil of generations of our ancestors, and we should value them highly.

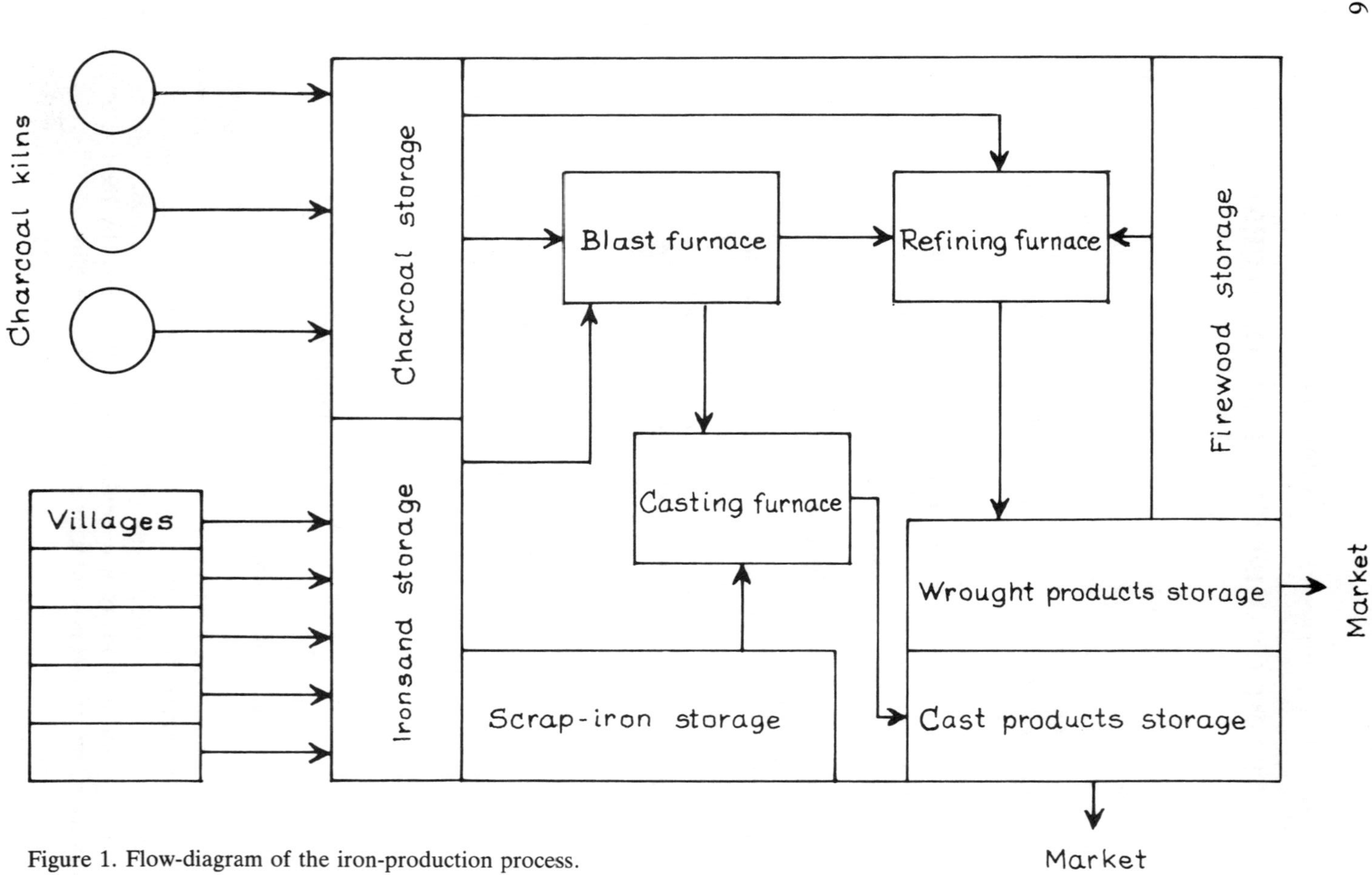

Figure 1. Flow-diagram of the iron-production process.

Figure 2. Ironsand sluicing at Lijiazhai, Xinyang County, Henan, about 1917 (Tegengren 1924, vol. 1, plate 16).

The local iron-production methods of Henan use ironsand and charcoal in a small blast furnace. The blast is provided by a manually-operated windbox.[3] This blast furnace produces white pig-iron, which is not suitable for working; therefore it is refined to make wrought-iron. This is done by breaking up the pig-iron and refining[4] it in a local type of low-temperature refining furnace. The fuel used is a mixture of wood and charcoal, and the blast is provided by a manually-operated windbox. The iron is then taken from the furnace and hammered to remove slag. The wrought-iron produced has a composition quite similar to that of mild steel; therefore the process is sometimes referred to as 'steelmaking'. Another term is 'puddling'.[5]

The Henan iron-production techniques use local materials and produce for local consumption. No special equipment or technical knowledge is required. Building an ironworks takes a little over two months, and an expenditure of several ten-thousands of yuan.[6] A single ironworks can produce several hundred tons of iron and steel per year; some produce over a thousand tons. Therefore the recommendation by the Central Committee of the Communist Party, that iron-production by local methods should be expanded, has a definite basis. There are, however, disadvantages to the local methods. For example, the production is low, the quality is inferior, a great deal of hard labour is required, and the cost to the consumer is high. These problems must, in the course of time, be overcome.

The special characteristics of the Henan iron-production techniques are:

- the ore is ironsand;
- the fuel is charcoal;
- no flux is used in smelting;
- the blast is cold;
- the blast furnace is very similar to a cupola furnace.[7]

These characteristics distinguish the Henan techniques both from modern techniques and from the traditional techniques of many other localities.

2.1.2. The production process

The Henan iron-production process, diagrammed in figure 1, is rather simple. It may be divided into the following stages.

Sluicing ironsand. Ironsand is found in large quantities throughout the Dabieshan, Tongbaishan, and Funiushan Mountains of southern Henan. Practically all the peasants of this region, young or old, male or female, know how to sluice ironsand. Generally one person can produce 40-50 kg of ironsand concentrate per day, in the peak season as much as 100-150 kg. It is the

Figure 3. Ironsand sluicing in Zhejiang, 1958 (Alley 1961b, no. 23).

principal sideline production of the peasants of the region. Figures 2 and 3 show the sluicing operation: the only equipment used is a wooden trough and an iron rake.

Producing charcoal. A great many kinds of trees and bushes grow thickly in the Dabieshan and Tongbaishan Mountains. The most suitable woods for charcoal production are teak and oak.[8] These are cut and charred in a local type of charcoal kiln. Each kiln is operated by 4-6 workers who specialize in this one operation. One kiln cycle requires 48 hours. The kiln is charged with 2000-2500 kg of wood, and 600-700 kg of charcoal are produced.

Smelting in a blast furnace. The local type of blast furnace is shown in figures 4 and 5. After the furnace has been thoroughly dried and preheated it

Figure 4. Blast furnace in Xinyang County, Henan, about 1917 (Tegengren 1924, vol. 2, plate 34).

Figure 5. Blast furnace in Xinyang County, Henan, about 1917 (Tegengren 1924, vol. 2, plate 34).

is charged alternately with ironsand and charcoal. The furnace is operated by shifts of three men each. In 24 hours 1400-1500 kg of ironsand are consumed, and 700-750 kg of white pig-iron are produced.

Refining.[9] The low-temperature refining furnace is operated by shifts of eight men each. In nine hours one furnace produces 500-600 kg of wrought-iron. In general the return is 85 kg of wrought-iron from 100 kg of pig-iron, consuming 20 kg of fuel (wood and charcoal).

The Henan iron industry supplies the needs of forty counties in Xinyang and Nanyang Prefectures, and also markets iron in the provinces of Hubei, Anhui, Hebei, and Shandong. In the past few years the applications for locally-produced iron have broadened considerably: white pig-iron is used in casting ploughshare-points, mouldboards, basins, woks,[10] etc.; wrought-iron is used for over 100 types of small implements, such as spades, hoes, sickles, mattocks, and axes.

A flow-diagram for the whole production process is given in figure 1.

2.2. The blast furnace

2.2.1. Raw materials

A modern blast furnace requires, in addition to ore and coke, flux and various supplementary materials. The Henan blast furnace requires only ironsand and charcoal.

2.2.1.1. Ironsand. Ironsand deposits are found in places where flowing water suddenly slows, as shown in figure 6. The ironsand is concentrated by sluicing; an analysis of the ore concentrate is as follows:[11]

SiO_2	CaO	Mn	P	S	FeO	Fe	Fe_2O_3	Fe_3O_4
5.52%	0.70%	0.29%	0.096%	0.011%	21.80%	64.55%	19.55%	70.24%

It can be seen that the iron content is very high, and the sulphur and phosphorous contents rather low. The iron content of the concentrate has a great effect on the smelting operation and its economics: thorough elimination of silica and other impurities is therefore very important.

2.2.1.2. Fuel. The local method of making charcoal, mentioned above, is very simple. It can be used anywhere, and requires no special equipment. The great disadvantage is that the very valuable by-products obtained in the distil-

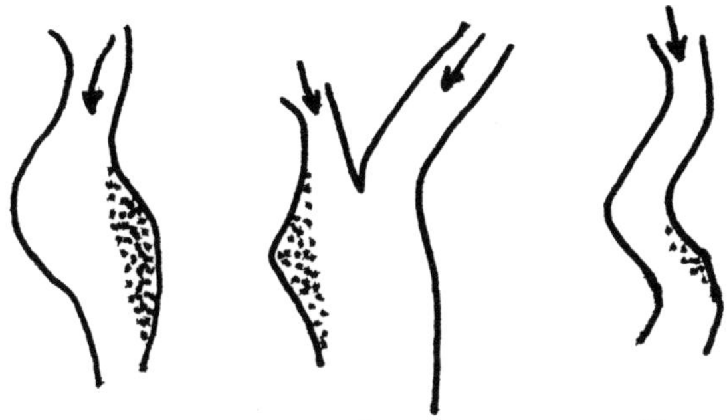

Figure 6. Ironsand placers: 1. Where a river widens. 2. Where a branch stream meets a major stream. 3. At a river bend. (Redrawn from a poor copy.)

lation process are lost. These by-products include coal-tar compounds and turpentine.

The yield of charcoal from wood is generally about 23 per cent, depending on the temperature of the kiln and the type of wood.[12] Because of the danger of spontaneous combustion it is best to store the charcoal in a dark dry place.

2.2.2. Blast furnace construction materials

The materials used in the construction of the blast furnace are as follows:

1. Furnace shell:
   30% loess, 70% quartz sand
   or
   20% loess, 10% rice straw, 70% quartz sand

2. Lining:
   30% powdered charcoal, 70% loess

3. Hearth:
   30% powdered charcoal, 60% loess, 10% gravel

4. Taphole stones:
   Two natural stones are placed above and below the taphole for resistance to heat and mechanical damage.

5. Shell:
   The furnace is reinforced with wrought-iron bands to prevent rupture.

6. Bottom:
   Three woks[13] are used to support the furnace.

7. Base:
   The entire weight of the furnace is supported by a wrought-iron ring with three legs.

8. Foundation:
   A large stone with three depressions for the legs of the furnace base.

2.2.3. Construction of the blast furnace

The blast furnace is diagrammed in figure 7. Its outer form is rather like that of a small cupola furnace,[14] but inside it is very different. The furnace is divided into three sections:

- The upper section is flared, and has a height of 1000 mm. This is the preheating zone.
- The middle section bulges on the outside. On the inside it bulges in front and is flat at the back. It is 1080 mm high. This is the fusion zone.
- The lower section consists of the hearth, the bottom, and the base. The hearth is a small hemisphere of refractory material supported by three woks and the tripod base. The height from the foundation is 240 mm.

The total height is 2320 mm, effective diameter 490 mm, effective height 1950 mm, inner throat diameter 940 mm. The inner diameter at the join between the upper and middle sections is 450 mm.

The most important characteristic of the furnace is that it is flared at the top. This increases the cross-sectional area where waste gases escape, so that the pressure is lower and the danger of flying charcoal and ironsand is reduced. The inside cross-section of the middle section is semi-circular; this makes possible the use of a single tuyère.

The fabrication of the furnace is straightforward. The tubular upper and middle sections are constructed as follows. A tubular basket is plaited of bamboo with the outside diameter equal to the intended inside diameter of the furnace. Then the prepared refractory material is spread uniformly on this form. The sections are placed in a dark place and very slowly blown dry; if they are exposed to strong sunlight they are liable to crack. Finally the furnace sections are bound with wrought-iron bands. Because of the slow drying, construction of a blast furnace requires 15-20 days, in some cases as long as 30 days.

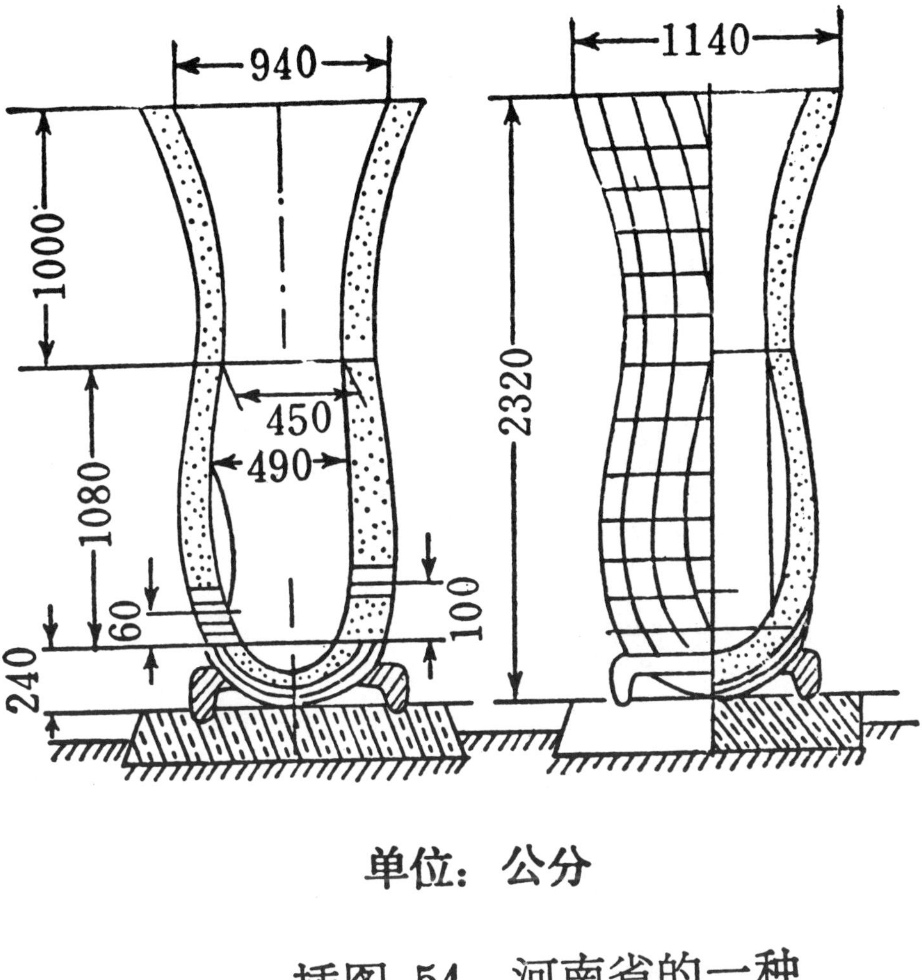

单位：公分

插图 54　河南省的一种
矮型土高炉
(采自河南省冶金局
编《土法炼铁》)

Figure 7. Diagram of the blast furnace. Reproduced by Yang Kuan (1982, p. 195) from the brochure translated here. Dimensions are given in mm.

2.2.4. Supplementary equipment for blast-furnace operation

item	*use*	*form*	*number*	*notes*
windbox	manually-operated blast	cylindrical	1	
wind-pipe	connection of the windbox to the furnace	hollow tube	1	50-100 spares
stopper-rod	sealing the taphole		2	
iron bar	tapping iron and removing slag	cylindrical	3	diam. 30-100× 1500 mm
shovel	charging ironsand		1	
enclosure[15]	tapping molten iron	semicircular shell	2	
platform balance	weighing ironsand and charcoal		1	
basket	charging charcoal		2	
pig mould	cooling the molten iron and forming it into a plate		1	
pond	cooling tools, breaking up slag, and fire prevention		1	

2.2.5. Operation of the blast furnace

In general the operation of the blast furnace is very similar to that of a cupola furnace. In the following only a few major points will be discussed.

2.2.5.1. Drying the furnace. A newly-built furnace contains a considerable amount of moisture; this can have a very deleterious effect on the operation and lifetime of the furnace, and it is essential that it be dried thoroughly.

All equipment should be carefully inspected before starting. Charcoal is charged in batches of *c.* 30 kg every 20 minutes. After ignition the blast may be started immediately. Drying a furnace requires 200-250 kg of charcoal, charged in 7-8 batches over 2-2.5 hours.

2.2.5.2. Blast. The furnace uses a manually-operated cold blast. Usually the windbox is operated by one person; if two persons operate it, production can be increased by about 20 per cent. However if the blast is too powerful there will be a danger of such deleterious effects as ‘flying sand’, ‘aggregation’, and ‘sinking charcoal’.

The dimensions of the windbox are: inner diameter 400 mm, effective piston-stroke 800 mm. The inner diameter of the blast-pipe is 100 mm. The windbox is operated at 20 strokes per minute. Thus with a utilization coefficient[16] of 0.80, the input of air to the blast furnace is

$$0.10 \text{ m}^3 \times 20 \text{ strokes/min} \times 0.80 = 1.6 \text{ m}^3\text{/min}$$

With an air density of 1.3 kg/m^3, this is

$$1.6 \text{ m}^3\text{/min} \times 1.3 \text{ kg/m}^3 = 2.08 \text{ kg/min}$$

The cross-sectional area of the blast pipe is 0.0078 m^2. Therefore the blast velocity is

$$(1.6 \text{ m}^3\text{/min}) \,/\, (0.0078 \text{ m}^2) = 205 \text{ m/min}$$

Since 1 kg of iron is produced every two minutes, the amount of air needed to produce 1 kg of iron is $2 \times 2.08 = 4.16$ kg. The total inputs required per kg of pig-iron are:

ironsand	2	kg
charcoal	3	kg
air	4.16	kg

2.2.5.3. Charge. When the furnace has been dried, and the charcoal has reached a red heat and is at a suitable height, charging of ironsand may begin. It is shovelled into the centre; care must be taken to avoid its being spread out. Immediately thereafter a measured quantity of charcoal is added. Each batch of ironsand is 20-25 kg, each batch of charcoal *c.* 30 kg. Batches are charged about every 20 minutes, always with the ironsand first and the charcoal second.

2.2.5.4. Tapping molten iron and slag. About an hour after blowing-in, molten iron will be seen dripping down.[17] After another half-hour slag is removed and molten iron is tipped out. In the first few tappings the temperature will be comparatively low and the quantity comparatively small. As operation

Figure 8. Tapping a blast furnace, Jinzhai County, Anhui, *c.* 1958 (*Ganze Volk* 1959). The entire furnace is tilted against the cross-beam.

Figure 9. Cartload of pig-iron, Dabieshan, Anhui, 1958 (Alley 1961b, no. 32).

becomes more regular, the temperature and the quantity of the iron will tend to stabilize.

The minimum temperature is 1200° C, the maximum 1400° C; normally it is in the range 1250-1300° C.

The minimum quantity of iron tapped is 15 kg, the maximum *c.* 30 kg; normally it is in the range 20-25 kg.

The molten iron is poured into a form and spread out with a wooden roller to form a flat plate which is easy to break for refining. Figure 8 shows the tapping of molten iron. [Figure 9 shows a cartload of these pig-iron plates.]

The chemical analysis of the pig-iron is:[18]

C	Si	Mn	P	S
4.2%	0.198%	0.257%	0.345%	0.027%

Slag is removed immediately before the iron is tapped. It is removed through the same taphole as the iron. The principal components of the slag are CaO and SiO_2. The temperature, viscosity, and chemical composition of the slag have a great effect on the quality of the pig-iron produced.

2.2.5.5. Troubles. Various problems can arise in blast furnace operation. The join between the upper and middle sections, where the diameter is smallest, is the part most exposed to damage. After the furnace has been in operation for some time the lining at this point can become uneven, so that the semifused ironsand can adhere to it. This impedes the descent of the charge, causing such effects as 'leaking sand', aggregates in the hearth, and incomplete separation of iron and slag. This may become so severe that operation must be halted. Local ironworkers call the phenomenon 'sand-sucking' or 'core-drawing'; the correct term is 'aggregation'. The aggregates adhering to the furnace lining should be pounded with a curved pointed iron rod until they break up and fall down. Then there will be aggregates in the hearth. This condition is corrected by raising the furnace temperature: more charcoal and less ironsand are charged, and the time between tappings is increased. If the aggregation is not extremely severe, furnace operation will quickly return to normal.

Another problem which can occur is 'canopying'. This can occur if the charcoal has a high moisture content, or is in too large pieces, or is very hard, so that it remains in the furnace for too long a time. When canopying occurs it will be seen that only the ironsand descends, while the charcoal does not. The effects of canopying are roughly the same as those of aggregation, and the means of correcting it are approximately the same.

It will be clear from the above that a blast-furnace operator must pay special attention to the manner in which the charge descends.

2.2.6. The blast-furnace smelting process

See figure 10. As the charge descends through the different temperature zones, in contact with the rising furnace gases, physical and chemical changes occur. The result of these changes is that molten pig-iron, *3,* drips downward, accumulates in the hearth, *7,* and is tapped through the taphole, *8.* The slag, *4,* formed from the gangue of the ironsand and the ash of the charcoal, drips down in the same way and collects in the hearth, *6.* Because of the weight difference it floats above the molten iron. It is tapped through the same taphole, *8.* . . .[19]

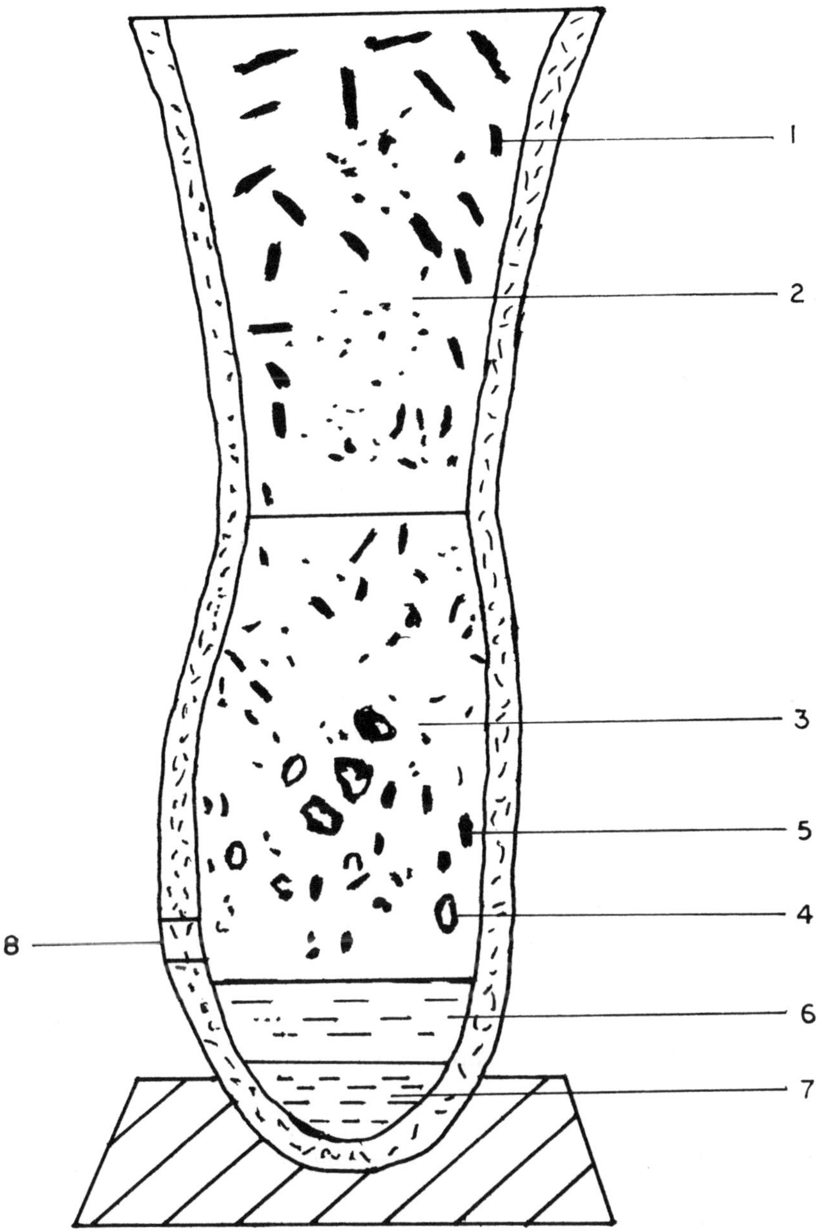

Figure 10. The blast furnace. (Redrawn from a poor copy.)

1. Charcoal. 2. Ironsand. 3. Semifused iron. 4. Drops of slag. 5. Drops of molten iron. 6. Molten slag. 7. Molten iron. 8. Taphole.

2.3. The refining furnace

As was mentioned above, the output of the blast furnace is white pig-iron; it can be used only in casting a few types of objects. Instead it is puddled in a local type of low-temperature refining furnace to remove carbon and some impurities to make a kind of mild steel. This can be used by smiths to produce all kinds of tools and agricultural implements.

2.3.1. Construction of the furnace; supplementary equipment

The furnace is diagrammed in figure 11. It is made of a mixture of loess and quartz sand, and thus it is an acid furnace. The blast-furnace pig-iron has very low sulphur and phosphorous contents, so that an acid furnace is suitable for refining it.

The furnace cavity, into which pieces of pig-iron are charged, is an oblate spheroid. The upper diameter is 228 mm, lower diameter 252 mm, middle diameter, at the widest point, 504 mm; height 306 mm. The total height of the furnace is 806 mm, total breadth 1200 mm.

The blast-hole in the cover is slanted, so that it is aimed precisely at the centre of the furnace cavity. The tuyère fits into the blast-hole; in order to increase the blast pressure, the tuyère narrows at the end.

Because of the violence of the puddling operation the furnace opening is made of three pieces of iron. The two vertical pieces each have one end buried in the ground, while the horizontal piece is secured by iron spikes into the ground.

The construction of the remaining parts of the furnace is not critical; they serve only to conserve heat and make operation more convenient.

The supplementary equipment required is:

- one windbox
- two anvils
- four flat-headed hammers
- two pairs of long-handled tongs
- 3-5 iron bars, diameter 30-40 mm, length 1500-1600 mm, for puddling
- two long-handled ladles, for removing slag

2.3.2. Operation

Figure 12 shows a refining furnace in use. Operation is fairly simple: it may be divided into the following steps.

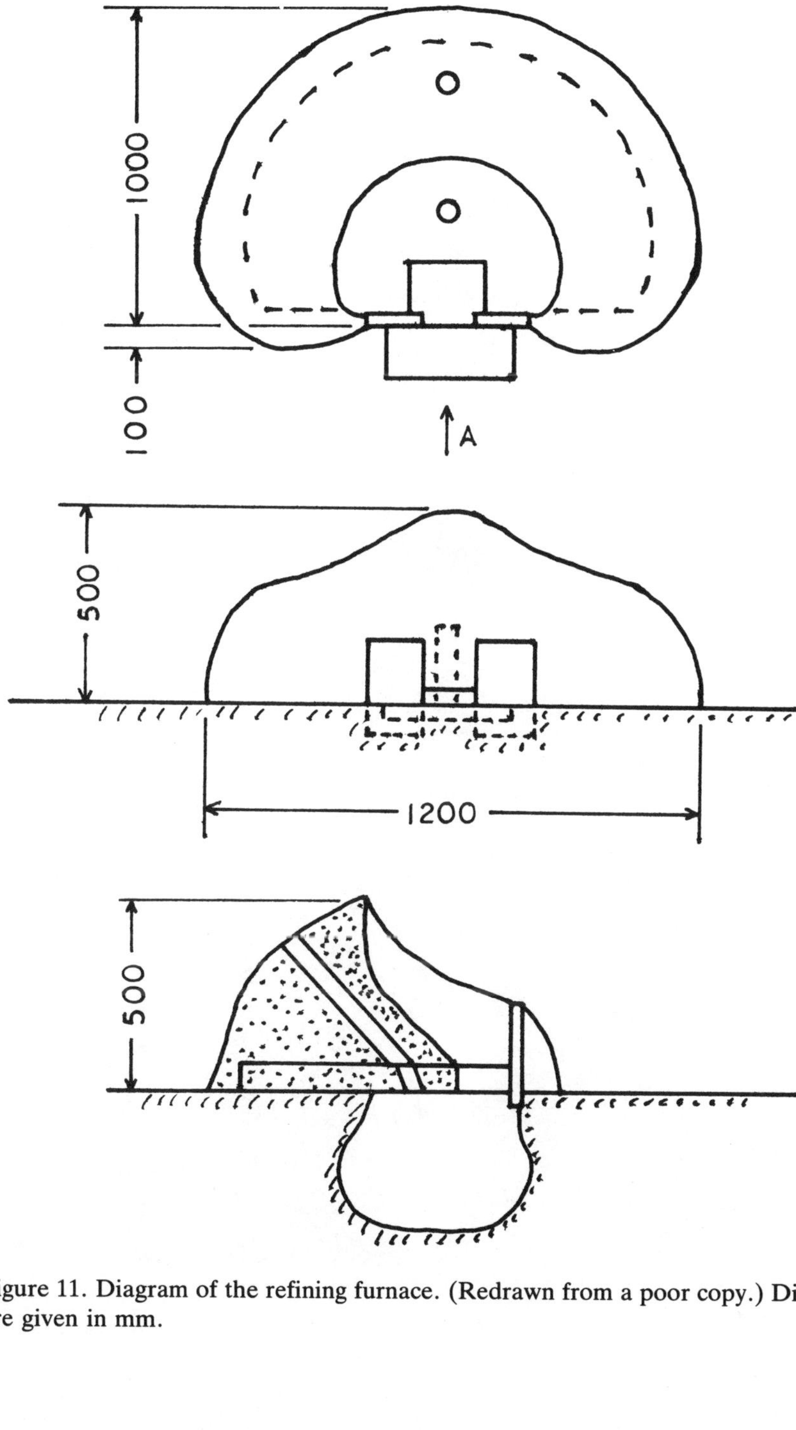

Figure 11. Diagram of the refining furnace. (Redrawn from a poor copy.) Dimensions are given in mm.

Figure 12. Operation of a traditional type of refining furnace in Shanxi, 1958 (Alley 1961a).

2.3.2.1. Charging. The dried furnace is charged with 7 kg of wood and 7 kg of charcoal. This is ignited, and the blast is started. Immediately 70 kg of broken-up pig-iron (each piece 20-30 mm) is charged, and the furnace opening is closed. The iron is covered with a layer of powdered charcoal, and refining begins. At this point red flames appear in the furnace.

2.3.2.2. Opening. After about twenty minutes white flames begin to appear; this indicates that it is time to begin puddling. The furnace opening is opened, and the operators begin violently puddling the iron. After about twenty minutes of puddling they begin removing the iron from the furnace and hammering it.

2.3.2.3. Discharging. The refined iron is removed from the furnace with tongs and hammered to squeeze out slag and compact the iron. The complete discharging of the iron takes about twenty minutes. Generally 30-40 wrought-iron bars are produced (see figure 13). Finally the hearth is cleared of slag with a ladle, and the cycle can begin again. Each cycle requires about sixty minutes.

2.3.3. Metallurgical conditions in the refining furnace

The purpose of any refining process is to lower the content of impurities to a certain level and to remove some of the carbon from the iron. In the process described here the sulphur content is reduced from 0.027% to 0.00324%, phosphorous from 0.345% to 0.206%, and carbon from 4.29% to 0.101%. The same holds for other elements.

In the local process the iron is never completely molten, but is in a semi-molten pasty state. Thus the temperature in the furnace is never very high, though at its maximum it can be as high as 1600° C.[20] As the iron is discharged from the furnace, and as impurities are removed from the iron, the temperature gradually decreases. When half of the iron has been discharged, the temperature is about 1400° C; thereafter it drops to 1350° C and finally to 1100° C. At this point the low temperature does not matter, as the refining process has already reached completion. . . .[21]

The analysis of the wrought-iron produced is:

C	Si	Mn	P	S
0.101%	0.096%	0.06%	0.206%	0.00324%

Figure 13. Bar iron produced by traditional methods in Hunan (*Ganze Volk* 1959).

2.3.4. Problems, furnace life, and economic indicators

The local refining method is so simple that in general no problems can occur; this is one of the great advantages of the method. Normally it is only necessary to assure that the furnace is airtight, that the blast is regular, and that puddling is started in time. If any problem develops, it will be due to one of these three factors.

Because the only materials used in the furnace are loess and sand, its heat-resistance and durability are limited, and it cannot be operated continuously. It should not be operated more than 8-10 hours per day, after which it should be cooled down; otherwise it will be ruined. The maximum life of a furnace is 20 days; in general it is 15 days.

The normal yield of wrought-iron from pig-iron is 85 per cent, the maximum 88 per cent. One furnace cycle requires about 60 minutes; the charge is 65-80 kg of pig-iron. Fuel consumption is not considerable, equal to about 20 per cent of the weight of the wrought-iron produced. Factors influencing the production rate are the quality of the pig-iron, the technical proficiency of the operators, and the construction of the furnace.[22]

CHAPTER 3

IRONSAND

Figure 14 shows roughly the areas of China in which to my knowledge there have, at one time or another, been iron industries based on ironsand. My data are undoubtedly incomplete, but these probably include the most important areas.

Ironsand seems to be a rather uneconomic ore: it can be seen in chapter 2, section 2.1.2, that production of ironsand concentrate per person per day ranges from 40 to 150 kg. The most primitive mining operations could probably produce more. On the other hand, ironsand-sluicing requires no special tools or skills, is outdoor work, requires less physical strength than mining, and involves no danger. Thus it is well suited as a sideline production for peasant families in isolated regions with certain characteristics: (1) there is a sufficiently large population to ensure a local demand for iron; (2) high transportation costs to outside markets make iron expensive and other local industries uncompetetive; (3) there is sufficient fuel for smelting. Besides these economic factors there are of course geological requirements; there must be large placers of ironsand, sufficient running water for sluicing, and a lack of easily-mined ore outcrops.

These factors seem to have been operative in most of the areas shown on the map, though those in Hebei province, close to the capital, are surprising. Of course a technical factor may also have been operative: ironsand iron may be superior to other types for certain purposes.

By the early twentieth century there seems not to have been much left of the ironsand iron industry, no doubt because of advances in railroads and shipping and the introduction of cheap iron, either foreign-made or made by

Figure 14 (opposite). Eastern China. Hatching indicates areas in which, to my knowledge, iron industries based on ironsand have operated at one time or another in the past.

Sources: General data: Nong shang 1919, p. 530; Tegengren 1924, pp. 96, 179, 207, 255, 259, 292, 307, 334, 336, 337; C.Y. Hsieh 1935, pp. 163, 182; Yang Dajin 1938, pp. 283, 287-9, 294-5. *Fujian:* Inouye 1912, p. 273; Yang Dajin 1938, p. 368. *Guangdong:* Li Longqian 1981, p. 365; C.Y. Hsieh 1935, p. 182 (only sources). *Hebei:* Zhu Xingzhong 1932, pp. 50, 64-5; Liu Yuncai 1978, p. 25; Sun & Sun 1966, p. 248. *Henan:* Guo Yujing 1932, p. 178; Zhang Youxian & Guo Yujing 1932, pp. 225, 234-7. *Hunan:* Yang Kuan 1960, p. 121 (only source). *Jiangxi: Nong shang* 1919, p. 530; Xia Xiangrong 1980, p. 234 (only sources). *Shanxi: Nong shang* 1919, p. 530; Sun & Sun 1966, p. 248 (only sources). *Shaanxi: Nong shang* 1919, p. 530 (only source). *Zhejiang.* von Richthofen 1872, pp. 47, 52; Yang Dajin 1938, pp. 313-15.

USSR
MONGOLIA
HEILONG JIANG
JILIN
LIAONING
KOREA
INNER MONGOLIA
BEIJING
Zunhua
TIANJIN
Mancheng
HEBEI
SHANXI
NINGXIA
QINGHAI
Linfen
Gaoping
Cixian
SHANDONG
Tengxian
Linyi
Cangshan
Zhengzhou
Gongxian
HENAN
Xi'an
SHAANXI
Wushan
Tianshui
Fangcheng
Nanyang
Tongbai
Xinyang
Guangshan
Shangcheng
Jinzhai
Hong'an
Macheng
Xishui
Qianshan
ANHUI
JIANGSU
Luhe
Nanjing
Shanghai
HUBEI
Yichang
Hanyang
Echeng
Jiayu
Daye
SICHUAN
Chongqing
Shaoxing
Ningbo
ZHEJIANG
Suichang
Longquan
Wenzhou
Changde
Changsha
HUNAN
Shaoyang
JIANGXI
GUIZHOU
FUJIAN
Fuzhou
Putian
Quanzhou
Shangyou
Gannan
TAIWAN
YUNNAN
GUANGXI
GUANGDONG
Guangzhou
Hong kong
VIET NAM
LAOS
HAINAN
0 90 180 270 360 450km
PHILIPPINES

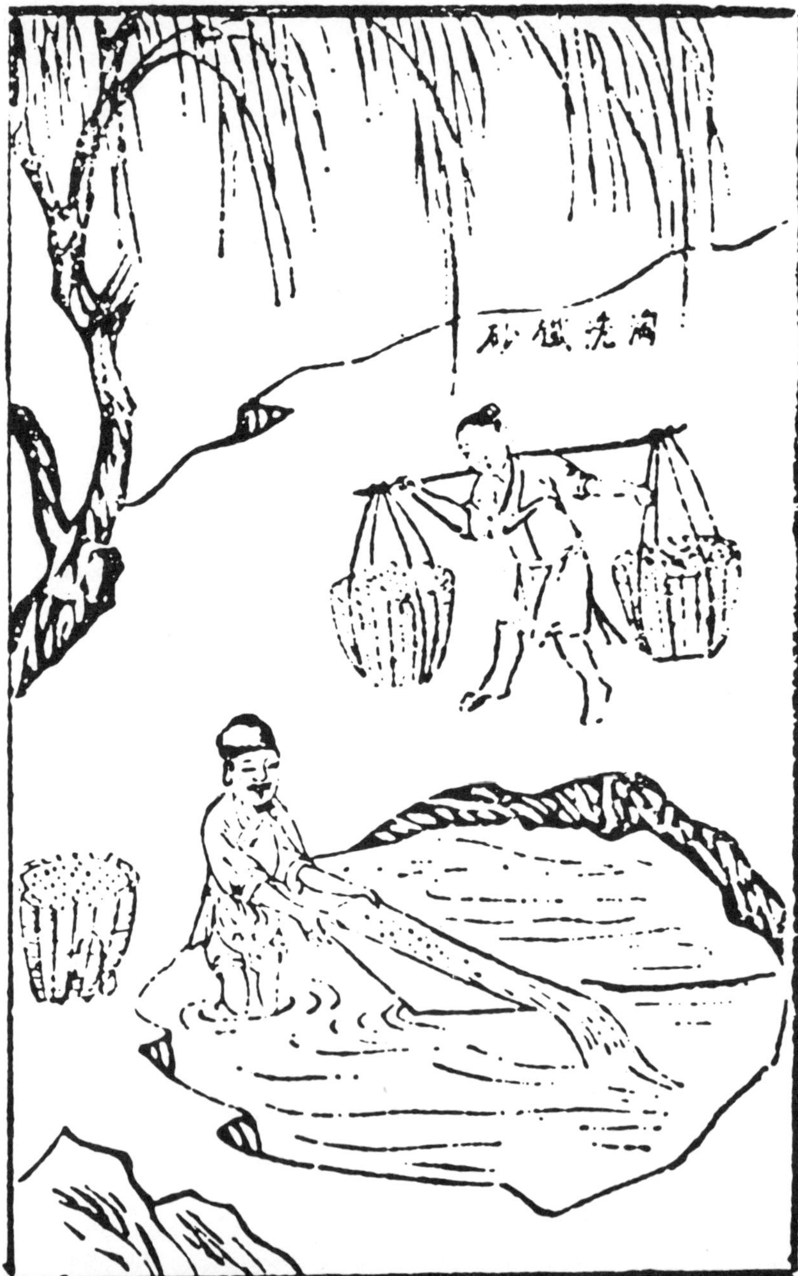

Figure 15. 'Washing ironsand', illustration from the early seventeenth-century technological encyclopedia *Tian gong kai wu* (TGKW, p. 362).

modern plants in China.[1] Southern Henan seems to have been the only part of China in which the ironsand iron industry was still flourishing in 1958.

The attempt to trace the history of ironsand smelting in China is frustrated by precisely the fact that it has been concentrated in isolated areas, for which there are very few sources. Xia Xiangrong (1980, pp. 234-6) quotes sources as early as the Tang period (A.D. 618-907) which indicate that iron was being smelted in areas which in more recent times have had ironsand iron industries. But none of the early sources (before the eighteenth century) quoted by Xia Xiangrong specifically mention ironsand, and it is quite possible that other ores were smelted in these places in earlier times. The earliest specific mention of ironsand smelting in China seems to be from the early sixteenth century, in Zunhua, Hebei.[2]

Another early notice of ironsand washing can be seen in the figure 15, from the early seventeenth-century technological treatise *Tian gong kai wu,* 'The products of Heaven and man'. There is an excellent translation of this book (Sun & Sun 1966).

It is startling that the picture shows *panning* in still water rather than *sluicing* in running water. The labour required must have been enormous – if we believe the picture. It is possible that the picture is an artist's conception, based on a knowledge of gold-panning rather than ironsand-sluicing.

On the method of ironsand sluicing used in southern Henan about 1917, Tegengren quotes Nyström as follows (see figure 2):

> The sand is dug out by means of iron scrapers and shovels and carried to the sluice board. 'This is built of wood, 1.6 m long and 0.5 m wide and provided with side boards 10 cm high, parallel or converging as the case may be. The board is set at an angle of 15° to the horizontal. A low embankment of earth is built on the two sides of the head, and the water from the river is led into the pool thus formed, from where it overflows down the board. While the sand is being spread over the board in a thick layer the water is barred with a short piece of board, but is then let on. While the water is quickly rushing over the sluice the sand is scraped with a pronged fork, which is moved up and down, and the result is that the lighter part of the sand is washed away and ultimately there remains an impure mass of magnetite and sand grains. Two or three men work together and there are no claims or permits in force. The total number of such ore washers in South Honan cannot be much less than 2500.'
>
> (Tegengren 1924, pp. 179-80)

Zhang Youxian and Guo Yujing (1932, p. 239) have another description of ironsand sluicing in southern Henan, and there is also a somewhat confused description of the same in Zhejiang (Yang Dajin 1938, p. 314).[3] Note also the information on ironsand sluicing in Northern Hubei quoted in chapter 5, section 5.5, below. The various descriptions differ slightly in the dimensions given for the sluices, but otherwise they are in general agreement. Both Tegengren and Yang give some information on prices and earnings, but interpreting these in real economic terms would require a good deal of research.

Ironsand was also the only ore used in the traditional Japanese *tatara* furnace (or 'horizontal blast furnace'). A film about *tatara* iron-smelting made by the Iron and Steel Institute of Japan shows briefly the process of ironsand sluicing used there: the sluices seem much larger and more elaborate than those described by Nyström.[4] According to the film the *tatara* process was last used in 1921. In this connection it is interesting to note that during the First World War ironsand was exported to Japan from Henan and Zhejiang (Tegengren 1924, p. 255; Guo Yujing 1932, p. 178). It would be interesting to know more about this. The Japanese importation of ironsand from China may have been a way of extending briefly the life of the vastly uneconomic *tatara* process, taking advantage of depressed conditions and low wages in China. On the other hand the ironsand may have been sintered and then smelted in modern blast furnaces.

CHAPTER 4

FORESTRY AND CHARCOAL PRODUCTION

It is remarkable to see how much charcoal was used by the traditional iron industry in southern Henan. What have the ecological effects been? The 1958 description (chapter 2, section 2.2.5.2) indicates that 3 kg of charcoal are consumed for each kg of pig-iron produced;[1] the yield of charcoal from wood is stated to be 23 per cent. Combining these figures with Nyström's estimate that 14400 tons of iron were produced per year in this region, it would appear that about 180000 tons of wood were consumed per year in iron-smelting alone. Percy (1875, p. 198) gives figures on annual forest growth between 2.5 and 3.4 tons of wood per hectare (10000 square metres) in well-managed forests in the Black Forest in Germany. It is really impermissible to combine figures from such diverse and uncertain sources as these, but if we do it anyway we can arrive at 500-700 square km as the area of well-managed forest needed to supply the fuel needs of this industry. This calculation cannot be used for any purpose except to give a rough impression of the order of magnitude of the problem.[2]

Even if the true figure were much larger it seems unlikely that the charcoal consumption of the iron industry would cause ecological devastation in the mountains of southern Henan – with the important condition that good forestry practices are maintained. Tegengren (1924, p. 334) indicates that they were:

> The charcoal is made of small fir-trees which grow in multitudinous patches on the foothills of the main range ... There are certainly millions of such trees, nearly all below 10 cm at breast height, which owe their existence to a settled system of ownership, protection, and rejuvenation. The wood is charred in numerous small kilns scattered amongst the hills.

Other observers have noted the existence of traditional forest management practices in various parts of China;[3] for example, in a study of the forest flora of Zhejiang, Anhui, and Jiangxi:

> It is comforting to find that the deforestation in southeastern China is not as bad as generally pictured. In many districts there are still magnificent forests, large tracts of them, indeed, properly cared for by the peasants, who usually understand the value of forests and look after them with an almost scientific correctness which invites admiration.
>
> (H. H. Hu 1929, pp. 55-6; see also p. 48)

The only detailed description of traditional Chinese forestry practices which I have found refers to northern Guangdong. Practices so far south obviously cannot be directly relevant to those of southern Henan, but they may be considered briefly here as an example of the level of sophistication which could be attained. They are an interesting combination of agriculture and forestry. A section of the forest plantation is harvested, burned over, and planted with upland rice or maize. The next year cuttings of Chinese fir (Cunninghamia lanceolata) are planted 2-3 m apart; between these a 'nurse crop' of cassava is planted. The cassava grows quickly and provides the shade needed by the young trees. After three years the shade is no longer needed, and the cassava roots are harvested. The trees are harvested after 30 or more years. Thus in a stable forest plantation there were three crops each year: grain, cassava, and timber (Pendleton 1937, pp. 474-5). Fenzel (1929, p. 93) considered this kind of forestry 'crude in its methods and narrow in its ends'; but Pendleton (1937, p. 478) noted that 'modern' forestry in this area was unsuccessful precisely because the traditional methods had been ignored.

A proper treatment of the earlier history of forest management in China would require a great deal of research. It may be hoped that the section on the subject in Joseph Needham's *Science and Civilisation in China* will be completed soon. Here I shall only mention a few interesting points.

Chinese thinkers were aware of the importance of proper forest management in very early times. Of many examples one is classic, the advice of the philosopher Mencius (372-289 B.C.) to King Hui of Liang:

> If you do not interfere with the busy seasons in the fields, then there will be more grain than the people can eat; if you do not allow nets with too fine a mesh to be used in large ponds, then there will be more fish and turtles than they can eat; *if hatchets and axes are permitted in the forests on the hills only in the proper seasons,* then there will be more timber than they can use. . . .
>
> (tr. D. C. Lau 1970, p. 51; italics mine)

The Chinese state took the initiative for large afforestation projects in A.D. 960 and in A.D. 1368 (the beginnings of the Song and Ming dynasties) (Chen Chi-yün 1939, p. 55), and probably on many other occasions.

Shiba (1970, p. 92) states that while logging was obviously a very old activity, the commercial cultivation of trees was new in the Song period (960-1279). In this period a magistrate in Quanzhou, Fujian issued a rhymed proclamation which can tell us something of the extent of deforestation, and a little of forestry practices, in Fujian in the Song period:

> The waste hills here aspire to reach the skies,
> Their soil as rich as flour, grass thick as felt.
> In the right season plant pine and cryptomeria seeds,
> And reap your reward a mere ten years hence.
> . . .
> There's no good reason not to grow them,
> But most men lack the foresight to plan ten years ahead.
> It fills me with regret to see
> Bare hills everywhere I look.
>
> (Shiba 1970, p. 97)

In early historical times Fujian was covered by continuous forest (Mell 1933). Agricultural settlement probably began here in the first centuries A.D., and it can be seen from the above that deforestation was already a problem in the Song period. In 1943 it was estimated that 18 per cent of Fujian's land area was forest and 31 per cent was bare land suitable for forestry; the remaining 51 per cent was agricultural land (D.Y. Lin 1943, p. 527). It seems that without public and private efforts at forest conservation over the centuries the province would have been as bare today as north China is. In the same study Henan was estimated to be only 0.6 per cent forest and 29.4 per cent bare land.

It is only in the past few centuries that China's explosive population growth and unstable political conditions have led to the thorough destruction of China's forests (Ho Ping-ti 1959, pp. 146-9; Wen Huanran & He Yeheng 1982). Modern efforts at reafforestation began in the early twentieth century, but made no significant progress until the establishment of the People's Republic. The problem is still very serious (Richardson 1966).

The translated description from 1958 indicates that charcoal was produced in southern Henan by specialized groups (households?) of 4-6 persons which could produce 600-700 kg of charcoal every two days. Thus there may have been 400 or more of such groups, each producing perhaps 100 tons of charcoal per year from a plantation of a few hundred hectares. Perhaps they combined forestry with some form of agriculture, as was done in northern Guangdong.

According to Tegengren (quoted above), Zhang Youxian (1932, p. 239), and the translated description (chapter 2, section 2.2, above), the charcoal is charred in kilns of some sort rather than in meilers (piles or stacks), as was normal in the European iron industry through the nineteenth century (Percy 1875, pp. 366-84). No other information is given, and the only description of traditional Chinese charcoal-production methods which I have been able to find is that of the Russian engineer M. Kovanko (1838).[4] Kovanko does not indicate where he observed the techniques he describes, but it was most likely in the neighbourhood of Beijing.[5] Figure 16 is his diagram of one type of kiln;

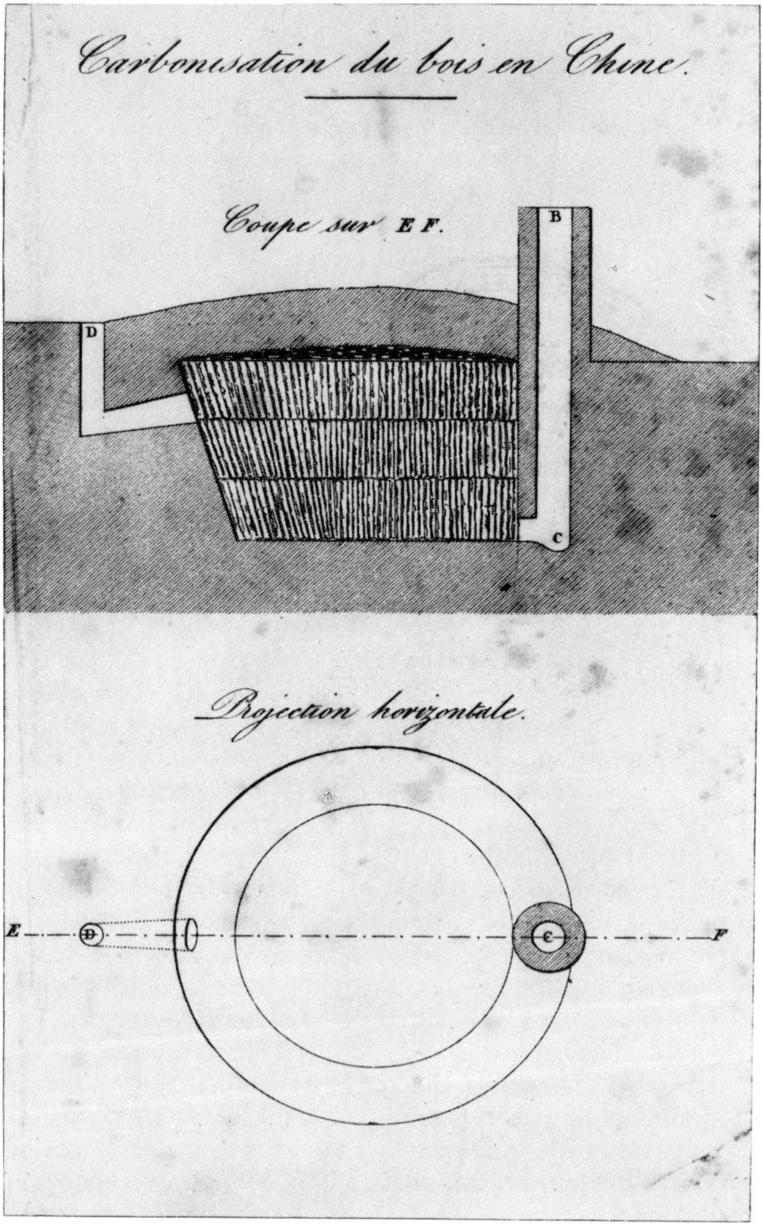

Figure 16. Charcoal kiln, probably near Beijing, *c.* 1838 (Kovanko 1838, plate II; see also Percy 1875, pp. 384-6). The depth is never more than 2½ *archines* (1.8 m); the diameter may be up to 2 *sagènes* (4.3 m).

it is possible that the charcoal kilns of southern Henan resemble this one. However his description indicates that its operation is rather different: the charring of one kiln-load requires 10-11 days as opposed to 2 days, and the yield is 30-35 per cent rather than 23 per cent.

The remains of old charcoal kilns near Hong Kong have been described briefly by James Hayes (1971), who also quotes some earlier accounts of charcoal-burning in that area.

Coal was used in China as the fuel in iron smelting as early as the Song period (960-1279), and probably much earlier (Hartwell 1962; 1966; 1967; Needham 1958, p. 19). In order to understand why charcoal was still being used in southern Henan, we must consider what information we can find on the relative costs of the two fuels. H. A. Little (1908, p. 12) lists the following prices for fuel in Yichang, Hubei, a small city on the Changjiang (Yangtze River), south of the Dabieshan region:

coal: 4d. for cheapest local to 1s.8d. per 133 lbs for best Sichuan.
firewood: 6d. to 8d. per 133 lbs.
charcoal: 'slightly dearer' than firewood.

Thus even in Yichang, a river port, charcoal could compete with coal; but these prices were probably very largely determined by the cost of transport: by river for coal, by coolie for firewood and charcoal. (This explains why charcoal was only slightly more expensive than firewood.) In southern Henan, close to the forested hills and separated by a mountain range from the Changjiang, charcoal would probably have been considerably cheaper than coal of adequate quality for blast-furnace use. The economic advantage of charcoal becomes even greater when it is considered that a flux would have been required if coal had been used. The economic advantage of charcoal disappears, of course, when consumption increases beyond the level of ecological equilibrium: a point which seems to have been overlooked by those who wished to expand local iron production in the 'Backyard Furnace Campaign'.

CHAPTER 5

BLAST FURNACES

In the description translated in chapter 2 the blast furnace is about 2 m high, flared at the mouth and with a narrow 'waist'. In tapping, the whole furnace is tilted. Furnaces of this type seem to have been used only in the Dabieshan region, but small furnaces with flared mouth seem to have had a wide distribution. The flared mouth appears to be essential in ironsand-smelting; the larger cross-section means reduced pressure, so that the ironsand is not blown out of the furnace by the top-gas. On the other hand, in some places flared furnaces were also used in smelting mined ore, charged in larger pieces. This was the case for example in Hunan. Another interesting aspect of the use of this type of furnace in Hunan is that the fuel used was coke or anthracite rather than charcoal (Tegengren 1924, p. 339; Lux 1912).

Sections 5.3-5.5 below consider three other descriptions, from 1917, 1932, and 1958, of waisted dwarf blast furnaces in the Dabieshan region. Section 5.6 takes up some technical questions on their operation. Before starting on this detailed discussion of one type of blast furnace we shall look briefly in the next two sections at a few of the many other traditional types of blast furnace which were in use in various parts of China in 1958, as well as some ancient blast furnaces.

5.1. Other traditional blast furnaces in twentieth-century China

The historian Yang Kuan, in an important monograph on the development of iron and steel technology in China, has studied the traditional blast furnaces used in the 'Backyard Furnace Campaign' (1960, pp. 117-56). He divides these into three classes, 'large', 'medium', and 'dwarf'. The blast furnace described in chapter 2 is among the larger of the dwarf types.

Figure 17 shows another dwarf blast furnace, called a 'trumpet furnace', used in Fujian. It is about 2 m tall, and essentially made of porcelain: the body is 98% kaolin (china-clay) and 2% paper-pulp or rice straw; it is lined with a mixture of kaolin and powdered charcoal. Unfortunately no information is given on its operation (Yang Kuan 1960, p. 119).

Figure 18 shows the same type of furnace in the Philippine Islands in 1902. No doubt it was introduced there by Chinese immigrants, perhaps from Fujian. The fuel used was charcoal. The ore, not mined but gathered from surface outcrops, was generally 60% iron. It was charged in pieces of about 1½ in (4 cm) diameter. No flux was used, and the loss of iron in the slag was about 20 per cent.[1]

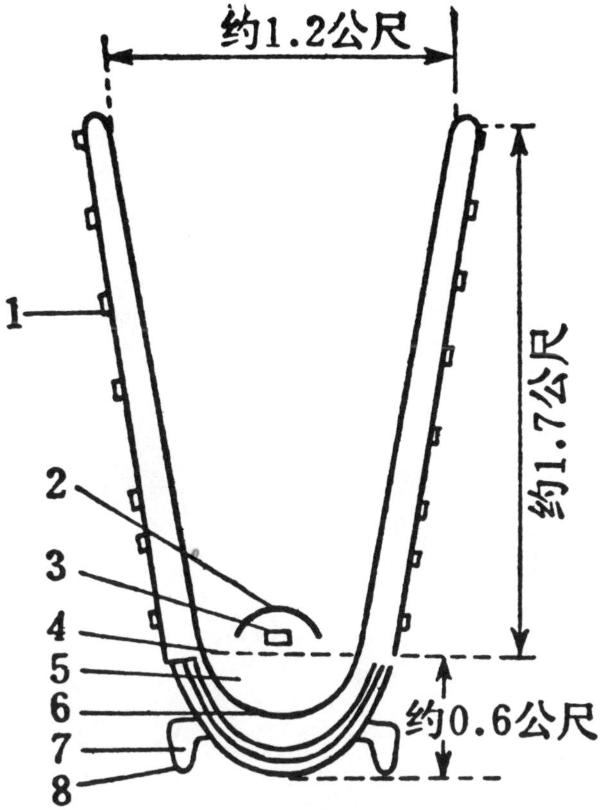

插图 53　福建省的一种"喇叭炉"
(采自陕西省科学技术协会筹备委员会编《土法炼铁》)
部位名称：1.扁铁　2.观音面　3.出渣口　4.出铁口　5.炉缸　6.炉底　7.铁锅　8.铁三脚

Figure 17. Diagram of a 'trumpet blast furnace' used in Fujian in 1958 (Yang Kuan 1982, p. 195). Total height 2.3 m, inside diameter at mouth 1.2 m.

1. Iron hoops. 2. 'Face of Avalokitesvara' (*Guanyin mian*) (?). 3. Taphole for slag. 4. Taphole for iron. 5. Hearth. 6. Bottom. 7. Iron woks. 8. Iron tripod.

Figure 18. Blast furnace in Bulacan, Philippine Islands, in 1902 (McCaskey 1903a, plate AJ; see also pp. 54-5 and plates VII-VIII). Height from ground 7 ft 5 in (226 cm), inside diameter at mouth 2 ft 8 in (81 cm).

Figure 19. Blast furnace in Wushan, Gansu, 1958 (Alley 1961b, no. 55).

Figure 20. Blast furnaces under construction in Wuwei, Gansu, 1958 (Alley 1961b, no. 69).

Figure 21. Blast furnace under construction in Shaoxing, Zhejiang, 1958 (Alley 1961b, no. 72).

Figure 22. Blast furnaces in western Hunan, 1958 (Alley 1961b, no. 10). Blast furnaces like these consist of an inner shaft of sandstone or brick, an outer wooden frame, and pounded sandy clay between the two (Yang Kuan 1960, p. 133).

Yang Kuan describes briefly several other traditional dwarf blast furnaces, some under 1 m in height. The most curious of these is a blast furnace about 30 cm (*thirty centimetres*) high; it resembles a water-bucket. The inside diameter is 3 *cun* (10 cm) at the top, 5 *cun* (16 cm) at the boshes, and 4 *cun* (13 cm) at the bottom. Production is normally 40-50 *jin* (20-25 kg) per day. This type of furnace was traditionally used in the region of Echeng County, Hubei, and apparently nowhere else. On the eve of China's national day, 1 October, in 1958 there were 62 544 of these furnaces in blast in Echeng County, with a total daily production of 1500 tons of pig-iron (Yang Kuan 1960, pp. 124-5).

The larger types of traditional blast furnace demand a separate detailed study. Here figures 19-22 will have to suffice for the moment: they show four of the largest traditional types of blast furnace used in China in 1958. Such furnaces ranged in height from 6 to 10 m (Yang Kuan 1960, p. 132).

5.2. Blast furnaces in ancient China

The earliest iron artifacts so far found in China date from about 500 B.C. These include both wrought-iron and cast-iron (Huang Zhangyue 1976). The wrought-iron appears to have been produced by the bloomery process (see chapter 6, section 6.2, below); but how was the cast-iron produced? No furnaces have been excavated from such early times, so here we can only guess.

I suggest that the first attempts at iron-smelting in China involved charging iron ore instead of copper ore in a small copper-smelting furnace.[2] Experience in copper smelting would have led the experimenters to expect a liquid metal. Using a stronger blast and larger amounts of fuel than in copper-smelting, it would have been possible to produce molten iron; Chinese bronze-casting techniques were so advanced at this time that the iron could be cast without much difficulty. Two things would soon have become apparent: (1) white cast-iron was an adequate material for some purposes, but poorly suited to many others; (2) lumps of what amounts to 'bloomery iron', probably steel with a fairly high carbon content, were left behind unmelted in the furnace, and these could be forged into very fine weapons and implements. Further experience with this new metal would in time have led to the development of two specialized types of furnace: the blast furnace and the bloomery. Wrought-iron from the bloomery was used for swords, knives, and other luxury items; cast-iron from the blast furnace was used for mass-produced implements and weapons such as shovels, hoes, ploughshares, axes, and arrowheads.[3] Techniques for refining pig-iron to wrought-iron, developed by the first century B.C., made the bloomery unnecessary, and it went quickly out of use (see chapter 6, section 6.2, below).

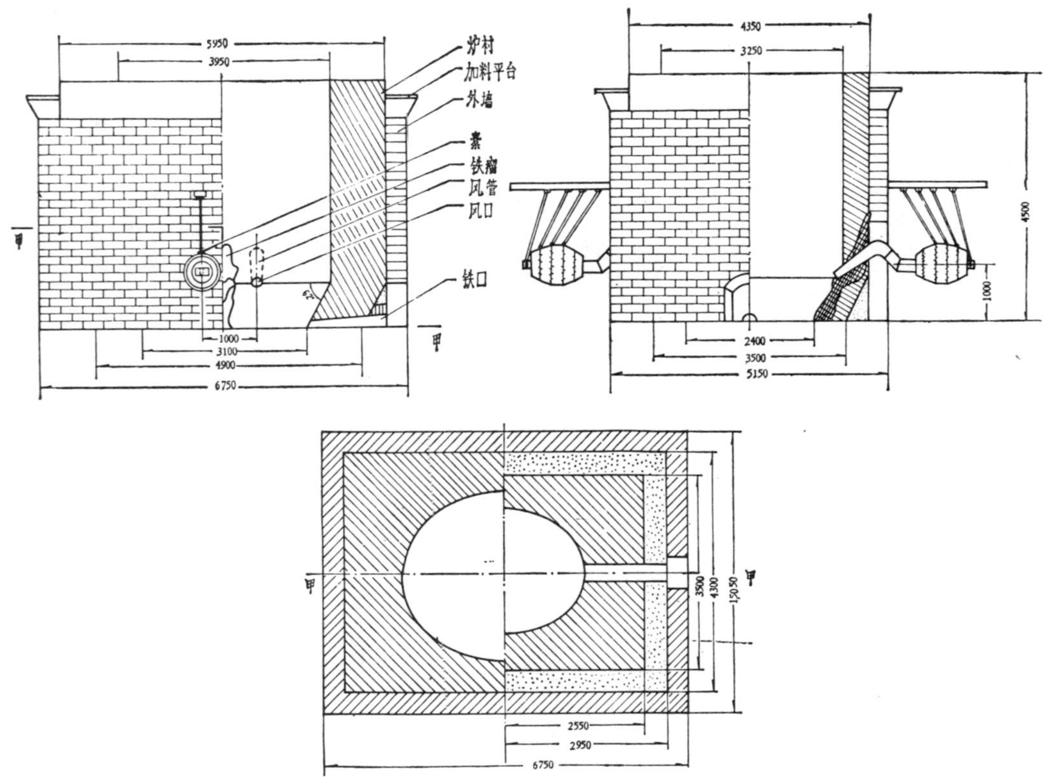

图三　古荥一号高炉复原图　（图中尺寸仅供参考）

Figure 23. Reconstruction of a blast furnace of the first century B.C., excavated at Guxingzhen, Zhengzhou, Henan (KGXB 1978.1: 8). Dimensions are given in mm. The height is shown as 4.5 m, but it may have been much greater. The bellows shown are reconstructed from a Han-period tomb-relief (see figures 43 and 44 below); I believe the actual blast apparatus used was probably much larger.

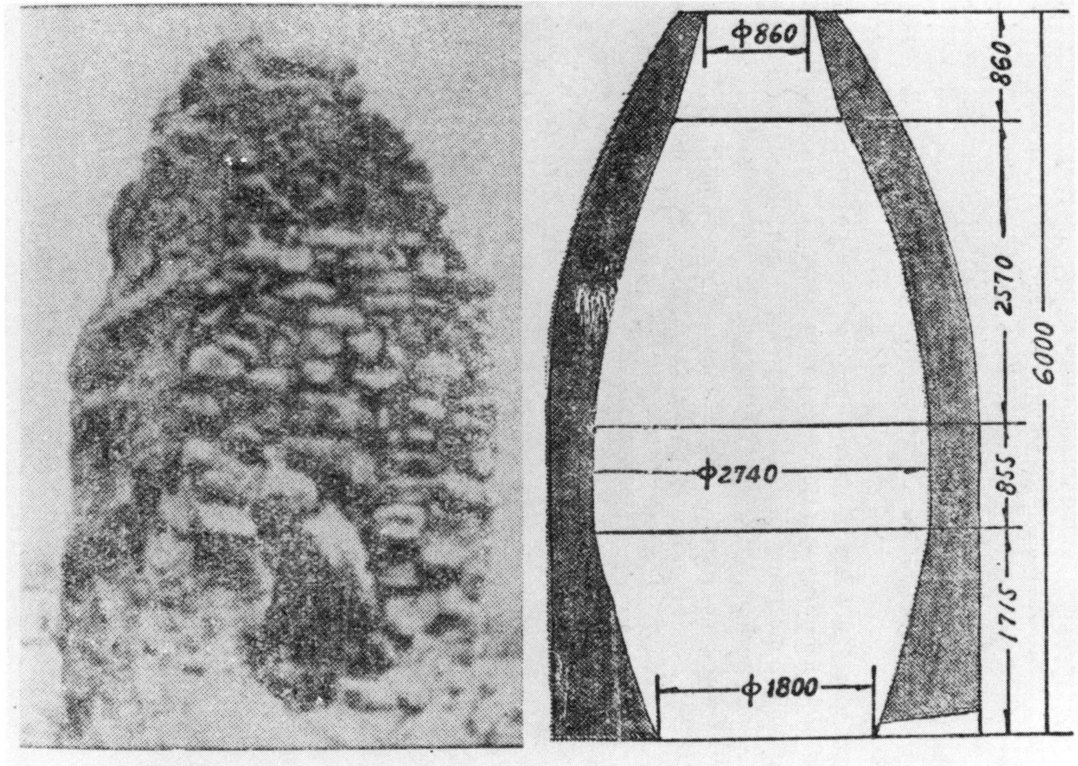

图八　矿山村宋代高炉　左：实物遗存　右：复原简式图

Figure 24. Blast furnace of the Song period (A.D. 960-1279) excavated in Hebei. *Left:* photograph of the furnace (reproduced by Liu Yuncai from Chen Yingqi 1959). *Right:* reconstruction by Liu Yuncai (1978, p. 23). Dimensions are given in mm.

The above discussion is speculative, since no remains of furnaces for iron-production have been excavated from before the first century B.C. Figure 23 shows a reconstruction of a blast furnace from this time, excavated at Guxingzhen, near Zhengzhou, Henan. The reconstruction is based on the well-preserved furnace foundation and a 20-ton piece of bear material which gives the shape of the hearth and part of the bosh. The hearth is elliptical, 3.1×2.4 m. The height has been variously estimated between 4.5 and 6 m. The ore used was a rich haematite, broken up and sieved so that the pieces of ore charged ranged in size from 2 to 5 cm. The fuel was charcoal, and limestone was used as a flux (KGXB 1978.1: 5-10; in English, Cheng Shih-po 1978; Tylecote 1983).

Chinese archaeological work on ancient iron has been largely concentrated on the earliest period, up to about the second century A.D. Surprisingly little has been published concerning the time after this. Figure 24 shows a blast furnace of the Song period (A.D. 960-1279), excavated in Hebei.[4]

A number of written descriptions of blast furnaces are extant from the time after the Han.[5] These will require a separate study; they cannot be dealt with here.

5.3. Nyström's 1917 description of a waisted dwarf blast furnace in southern Henan

The description by Nyström in 1917 and Guo Yujing in 1932 are rather brief compared to the 1958 description translated in chapter 2, but they add some useful details and show some variations in furnace construction and operation.

Tegengren's summary of Nyström's report gives the following description of the blast furnace and its operation (see figures 4-5):

> The blast-furnaces have an exterior shape reminding of a half barrel put on top of a whole one, the whole height being only about 2.5 m and the width about 1 m. The furnace is built of fire bricks and strengthened with longitudinal and latitudinal iron hoops. There is one opening at the back for the tuyère and one tap-hole in front for iron and slag.
>
> The blower consists of a hollow tree trunk, 2.5 m long and of 0.25 m inner diameter, with a piston operated by a long wooden rod. It is connected with the tuyère by pipes of fireclay, 6 cm inside diameter and 0.5 m long.
>
> The blast-furnace is at first charged with charcoal which is set on fire, whereupon the blast is started and then ore concentrates gradually added. One furnace is capable of turning out about 1000 catties (0.6 ton) in 24 hours.
>
> The molten raw iron running from the furnace is allowed to spread out in shallow moulds, thus forming thin sheets about 50×12×1 cm.

The raw iron sells at the furnace at from 40 to 50 cash per catty, or $50 to 60 per ton.
(Tegengren 1924, p. 335)

This is quite clearly the same type of furnace as that of the 1958 description. The statement that the furnace is built of firebrick is probably a mistake: Nyström's photographs (figures 4 and 5) show no sign of brickwork.

A major difference in operation is shown by the analysis (dry) of the ironsand (p. 180):

Fe	SiO_2	P	S
48.88%	25.84%	0.232%	0.052%

The ironsand is obviously much less concentrated than in the 1958 description: it contains almost five times as much silica. Tegengren implies (p. 335) that a limestone flux was used; this would surely be necessary with so much silica in the furnace charge.

Tegengren also states that the moist ironsand contains about 40% iron (p. 180); this figure, together with his production figures (p. 334 fn.), suggests that the ironsand was charged wet, with about 18% water. The 1958 description says nothing about drying the ironsand. If drying were important it would surely have been mentioned, so it seems likely that here too the ironsand was charged wet. We shall see in section 5 below that this water may have had an important function in spreading out the ironsand grains in the upper part of the furnace.

Nyström states that the furnace charge is three parts charcoal to one part ironsand (p. 334). Thus the inputs to produce one kg of pig-iron would be approximately

ironsand	2.5 kg
charcoal	7.5 kg
limestone	??

Tegengren considers this charcoal consumption 'incredibly high', and notes that it is ten times the amount required in a Swedish charcoal blast furnace. It is also 2.5 times higher than that given in the 1958 description; I suspect some mistake on the part of Nyström, but let us note that it is impossible to be sure on a point like this.

5.4. *Guo Yujing's 1932 description*

Guo Yujing describes a blast furnace in southern Henan in considerably more detail. For some reason he gives the measurements in English feet and inches, which seems odd for a Chinese writer:

The blast furnace, or 'sand-smelting furnace', has the following dimensions:

height 6 to 7-8 ft [183 to 213-244 cm]
upper diameter 2 ft 2 in to 3 ft 5 in [66-104 cm]
middle diameter 1 ft 2 in to 2 ft 10 in [36-86 cm]
lower diameter 1 ft 4 in to 3 ft 1 in [41-94 cm]

It is divided into an upper and a lower section. The outside is hooped with iron bands, four each on the upper and lower sections. These are joined together by several straight iron rods.

The back wall of the lower section is 9 in [23 cm] thick, and the front wall is 4 in [10 cm] thick. The upper section is 4 in [10 cm] thick throughout.

The bottom part is called the 'ocean bottom'. It is made of a large iron wok.

There is a hole with diameter 4 in [10 cm] in the back wall for the blast-pipe. In the front wall is a hole, *c.* 1½ in [4 cm] square, for tapping molten iron and slag.

The area around the taphole is most subject to melting. In Xinyang, sandstone from Jiayu County, Hubei, is used [for the taphole stones]; in Shangcheng, diatomaceous earth from Qishui [modern Xishui], Hubei, is used. The two materials are equally refractory.

Materials for furnace construction: Loess and sand are used, together with unburnt charcoal from furnace slag, washed and powdered. This is called 'charcoal powder' (*tanfen*). The loess is mixed with the sand and charcoal powder and plastered evenly on the inside of the furnace. The proportion of loess to white sand is *c.* 100 to 34. When this has dried, charcoal powder is plastered over the whole; then the furnace is ready. At the bottom, where the temperature is highest, old taphole stones (diatomaceous earth) and a small amount of salt are added to the charcoal powder, loess, and sand.

Tuyères: The material used is the same as above. This is formed into a long tube, 2 ft [61 cm] long, with outer diameter 4 in [10 cm], inner diameter 2 in [5 cm], weight *c.* 40 *jin* [*c.* 20 kg]. One furnace uses four tuyères every 24 hours.

The most important detail in the installation of the tuyère is its angle: the blast should be directed precisely toward the taphole. Otherwise the temperature will be insufficient.

Windbox: A wooden windbox is used for the blast. It is 7 ft [213 cm] long, with inner diameter 1 ft [30 cm], outer diameter 1 ft 4 in [41 cm]. Valves are installed inside. The shaft is 12 ft [366 cm] long. It is worked by the strength of two persons.

Blowing-in: First the furnace is filled with charcoal. This is burned until the furnace bottom is extremely hot. Then another layer of whole charcoal is added, followed by a layer of crushed charcoal, to a thickness of 2 ft [61 cm]. Then a smaller amount of ironsand is added, thickness *c.* 1 in [2.5 cm].

The blast is operated until the top of the furnace burden sinks down; then charcoal and crushed charcoal, thickness 5 in [13 cm], and ironsand, thickness 1 in [2.5 cm] are added continuously in a cycle. After the third addition of ironsand molten iron appears. Charcoal and ironsand are added one more time, and then the molten iron is tipped out. At the same time a hooked measuring rod is inserted into the tuyère. If the tuyère has been shortened by melting it is pushed further in, so that it is flush with the inner wall.

In one day and night the inputs are:

charcoal 4200 large *jin* [*c.* = kg]
ironsand 3000 large *jin*

and the output is:

pig-iron 1300 large *jin*

The molten iron is tipped into a large iron ladle which is lined with a mixture of salt and loess and further with charcoal powder. The iron is poured into a sand form 1 ft 8 in [51 cm] long and 5 in [13 cm] wide.

The furnace lining is rich in silicon, and this is not neutralized by an alkali. Therefore it is subject to corrosion. It should be repaired every two or three days.

The smelting slag does not flow [i.e. flow freely?]; it contains small pieces of charcoal and bean-shaped iron aggregates. These are crushed, washed, and returned to the furnace.

(Zhang Youxian & Guo Yujing 1932, pp. 239-41)

An analysis of the ironsand is given, but this analysis is identical to that given by Tegengren, quoted above (section 5.3); it is unlikely to be relevant here. No analysis is given for the pig-iron produced.

From the input and output figures given we may calculate that the inputs needed to produce one kg of pig-iron are:

ironsand	2.3 kg
charcoal	3.2 kg

These figures are very close to those given in the 1958 description (chapter 2, section 2.2.5.2).

5.5. A technical study of the operation of the Huang Jiguang blast furnace in northern Hubei

A similar blast furnace was the subject of a technical study by two workers from the Chinese Iron and Steel Institute (Liu Zhichao & Tang Youyu 1959). This was the Huang Jiguang Furnace[6] at the East Wind People's Commune in Macheng County, Hubei, on the other side of the Dabieshan Mountains from southern Henan. The furnace is diagrammed in figure 25. It is very similar to the one described in chapter 2 above, except that the mouth is not as widely flared. Its operation is rather different, using a much greater blast and much less charcoal.

The article gives a good deal of interesting technical information such as temperatures, gas pressures, and analyses. Unfortunately most of this information relates not to normal operation but to a special demonstration in which the furnace was pushed to a production rate probably several times higher than normal. An earlier study on this same furnace[7] is referred to, and this may include studies of normal furnace operation, but I have not been able to obtain a copy of it. The article starts with a brief description of the blast furnace and its normal operation. The only major difference in construction is that the material of the furnace lining is 70% charcoal powder, as opposed to only 30% in the blast furnace described in chapter 2 (section 2.2.2). This material should presumably withstand higher temperatures.

The most important detail in the furnace construction is that the tuyère should be aimed precisely at the upper taphole stone. This, according to technical studies and the experience of older workers, causes the blast to be divided into three parts: (1) up the stack; (2) out through the taphole (?); and (3) down to the hearth and then up the stack.

The normal charging sequence is: (1) crushed charcoal, spread out and packed down with a shovel; (2) ironsand, charged into the centre of the furnace mouth; (3) whole charcoal, length 15-25 cm. The amount used per charge is not indicated.

The windbox used for the blast is almost twice as large as that described in chapter 2 (section 2.2.5.2). the inside diameter is 38 cm and the effective piston stroke is 1.5 m.

A good deal of information is given on ironsand washing. The river sand is 2-7% ironsand; after the first sluicing it contains 61-63% iron. Experiments show that repeated washing can improve the iron content somewhat, but with considerable loss of material. Improvement is slight after the second washing. It appears that the ironsand is charged wet.

Analyses are given for four kinds of charcoal:[8]

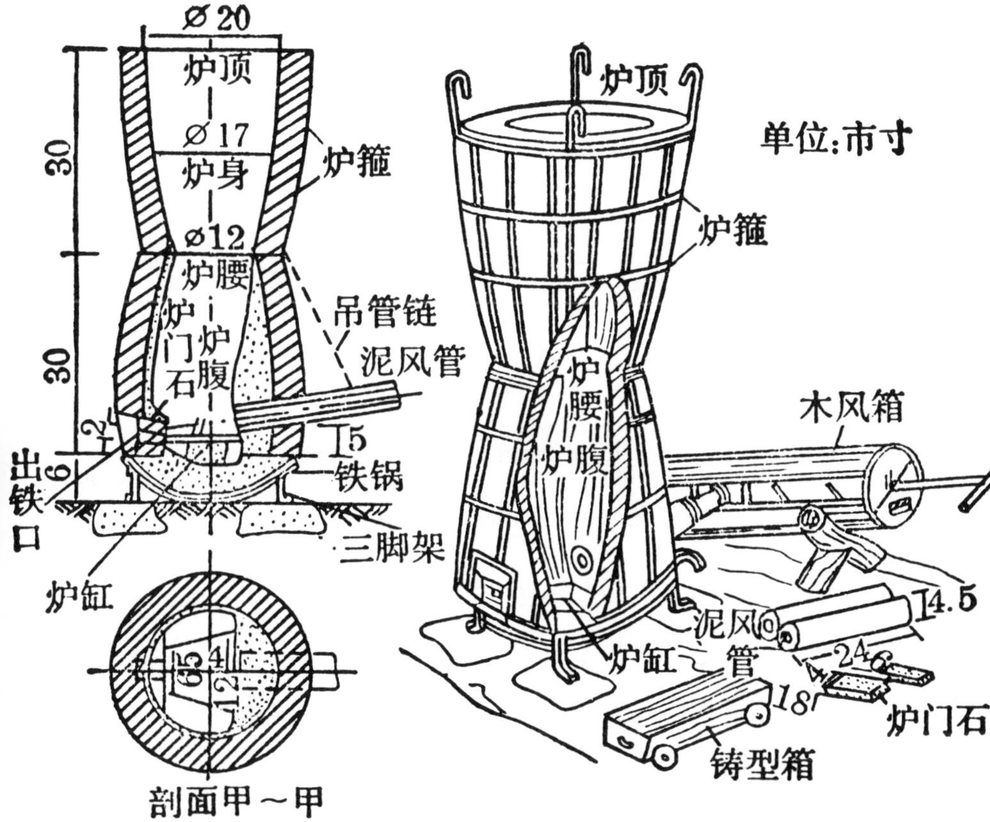

插图 52 湖北省麻城县的甑炉
(采自«冶金报»1958 年第 45 期)

Figure 25. Diagram of the Huang Jiguang blast furnace at the East Wind People's Commune, Macheng County, Hubei (Yang Kuan 1982, p. 193). Dimensions are given in 'market inches' (*shicun,* 3.3 cm).

	carbon	volatiles	ash	water
Chestnut white charcoal	83.5%	4.93%	5.41%	6.31%
Chestnut black charcoal	77.30%	15.24%	4.52%	2.49%
'Mixed wood' white charcoal	78.35%	5.59%	6.35%	9.71%
'Mixed wood' black charcoal	78.36%	16.89%	1.44%	3.31%

A special demonstration (called a 'Sputnik launch') was held on 13 January 1959 to investigate how high the production rate of the furnace could be pushed. Because all hands were needed in the fields at this time, the demonstration lasted only 2 hours and 27 minutes, from 1240 to 1507. The furnace appears to have been in normal operation both before and after the demonstration.

The charcoal and ironsand were analysed as follows:

> Charcoal: carbon 77.3%, volatiles 15.24%, ash 4.52%, water 2.94% (total 100.00%). This was chestnut black charcoal, used because white charcoal was not available at the time.
>
> Ironsand: Fe 62.12%, SiO_2 3.99%.

The distribution of diameters of ironsand grains is given as:

diameter (mm)	>0.43	0.43-0.25	0.25-0.15	0.15-0.09	<0.09
%	10.2	60.8	19.6	7	2.4

(Total 100.00%).

The windbox was operated by four teams of six persons each, changing every 1-2 minutes; obviously this was exhausting work. The stroke rate was counted nine times during the demonstration; it varied between 55 and 66 strokes/min, with an average of 61. In each cycle of two strokes the workers took two steps forward and three steps back. The blast was continuous except for the time required for tapping and changing tuyères, a total of 12 minutes.

Measurement of the blast pressure gave the curve in figure 26. The pressure was highest, 280-300 mm water, at the end of the forward stroke, and lowest, 20-40 mm water, at the pause between the forward and backward strokes.

It appears that this large variation in blast pressure, with a cycle of about 2 seconds, is important for the operation of the furnace. The authors cite experience with the use of a 2 HP electric blower on a blast furnace similar to this one. The blast pressure was fairly constant, between 100 and 120 mm

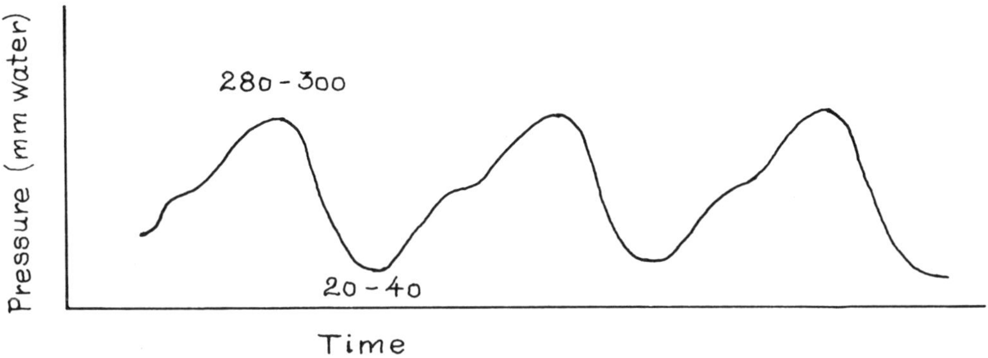

Figure 26. Variation of blast pressure in the stroke cycle of the windbox on the Huang Jiguang furnace (Liu Zhichao & Tang Youyu 1959, p. 182).

water; this is considerably below even the average pressure of figure 26, but there were problems with ironsand being blown out of the furnace mouth. 'Flying sand' was not nearly as much a problem in the Huang Jiguang blast furnace, and this is probably in some way related to the cyclic variation of its blast pressure.

We may also remember here the statement in chapter 2 (section 2.2.5.2) that the blast is normally worked by one man, and that if two men work it there is a danger of 'flying sand' caused by excess pressure. It appears as if the use of firmly-packed crushed charcoal as part of the charge in the Huang Jiguang furnace (and in the one described by Guo Yujing, section 5.4 above) may have the function of reducing this problem in spite of higher blast pressure.

The pressure of the escaping gas at the furnace mouth was measured as 10-30 mm water. (I wonder how this measurement was made.) The temperature of the gas was *c.* 370° C, and analyses of the gas gave the following results:

	CO	CO_2	O_2
At edge	32%	1.4%	0.2%
Halfway to centre	29%	1.6%	
At centre	24.2%	1.8%	0.8%

Temperature measurements were made through two holes in the furnace wall 15 cm above and below the 'waist', the narrowest part of the furnace. The temperatures were respectively 740° C and 1150° C. Attempts to take samples through these holes were unsuccessful.

The temperature in the hearth was 1570° C, and that of the slag, measured through the taphole, was 1400° C.

In the demonstration a large amount of scrap-iron was added to the charge; the authors state that this is not done in normal operation. The total charge during the 147 minutes of the demonstration was:

charcoal	435 kg
ironsand	575 kg
scrap-iron	316.5 kg

This was charged in 19 batches (i.e. every 8 minutes on average), and the iron and slag were tapped 8 times (every 18 mins). The total outputs were:

pig-iron	645.75 kg
slag	32.5 kg

Analyses of three samples of pig-iron were:

	C %	Si %	P %	S %
1	3.65	0.13	0.59	0.042
2	3.9	0.06	0.26	0.019
3	3.02	0.01	0.42	0.044

Analysis of the slag:

Al_2O_3	SiO_2	CaO	FeO	total
14.07%	26.24%	25.88%	16.1%	82.29%

Using the assumption that the scrap was 80% iron, the authors calculate that the amount of new iron produced was $645.75 - 0.80 \times 316.5 = 392.55$ kg. Inputs required to produce 1 kg of this new iron were thus:

charcoal	1.1 kg
ironsand	1.5 kg

This charcoal consumption is stated to be about the same as in normal operation; it is about one-third of the charcoal consumption in chapter 2 (section 2.2.5.2). It can be calculated that the iron lost in the slag was only about 0.2% of the total iron output.[9]

The authors also calculate that the air input was about 9 m^3/min. The blast was worked for 135 min of the demonstration period; thus we can calculate that the total air input in the blast was *c.* 1580 kg.

The authors note that they at first had assumed that each charge of ironsand sinters together into a mass in the furnace. Observation showed that this is not the case. They conclude that the water content of the ironsand, when it vaporizes, spreads out the grains so that they descend separately. In a study of another blast furnace, similar to this one, samples taken through a hole in the furnace wall 50 cm above the taphole contained fine particles of metallic iron, each presumably produced by the reduction of a single grain of ironsand.

5.6. *Technical notes on the waisted dwarf blast furnace*

We have now considered four descriptions of the construction and operation of this type of blast furnace. These give a great deal of technical information which will repay careful study; the following notes take up a few points which have especially attracted my attention.

In a modern blast furnace ironsand would never be charged without a preliminary sintering into larger lumps. The fine grains of ironsand would give two problems: they would be blown out the furnace mouth, and they would reduce the permeability of the furnace burden. The problem of 'flying sand' seems to be reduced in the waisted dwarf blast furnace by the flared furnace mouth. In addition, in the furnaces described in sections 5.4 and 5.5 above, the use of layers of crushed charcoal, tightly packed down, as part of the charge appears to help somehow with this problem.

In the Huang Jiguang furnace (section 5.5 above) it appears that the problem of the reduced permeability of the furnace burden is solved by the use of an extremely powerful blast, much more powerful than in the other three descriptions.

In the other cases the problem of burden permeability is solved by using a very large excess of charcoal in the charge. The excess is not consumed, but leaves the furnace in the slag. This can be shown as follows.

In Guo Yujing's description (section 5.4 above) it is stated that the slag contains charcoal, and that this is crushed, washed, and returned to the furnace. From the data given in chapter 2 (section 2.2.5.2) we can make a rough estimate of how much carbon the slag contains. The total inputs to produce 1 kg of pig-iron are given as 3 kg charcoal, 2 kg ironsand (wet), and 4 kg air. Rough calculations give an oxygen input of 0.9 kg in the blast air plus 0.4 kg in iron oxides in the ironsand, for a total of 1.3 kg.[10] The charcoal is probably about 80% carbon (section 5.5 above), so the total carbon input is about 2.4 kg. A negligible amount of carbon (*c.* 40 g) leaves the furnace in the pig-iron; the rest must leave either in combination with oxygen (CO or CO_2) in the top gas, or as elemental carbon in the slag.

Only about 1 kg of carbon can be oxidized to CO by 1.3 kg of oxygen. The rest of the carbon, about 1.4 kg (probably more), must leave in the slag.[11]

Since the ironsand contains very little gangue, and no flux is used, the other components of the slag (SiO_2, Al_2O_3, CaO, etc.) should not amount to much: by a very rough calculation, perhaps 200 g per kg of pig-iron produced. Thus the slag might be about 90% elemental carbon by weight. The percentage by volume would be even higher.

The slag presumably consists of small pieces of carbon bound together by the fused component. The description translated in chapter 2 (section 2.2.4) mentions a pond, used, among other things, for 'breaking up slag'. Pouring the slag into water should prevent most of the carbon from burning away upon contact with air.

It is odd that the translated description does not mention that the slag contains carbon which can be returned to the furnace. Perhaps the authors felt that this was a complication which should be avoided in a manual for beginners.

The special demonstration of the Huang Jiguang furnace (section 5.5 above) used a much more powerful blast and much less charcoal input. Repeating the above calculations using the data given for this furnace, it turns out that the oxygen and carbon inputs are approximately in balance, and no carbon would be expected in the slag. This inference is confirmed by the analysis given for the slag, and also by the very small total weight of the slag produced in the demonstration (32.5 kg slag for 393 kg new pig-iron produced).

The only figures we have on the operation of the Huang Jiguang furnace are for extraordinary conditions during a special demonstration. It is an interesting exercise to attempt to calculate what operation might have been like under normal conditions. It is stated that the charcoal consumption of 1.1 kg per kg new pig-iron during the demonstration was about the same as in normal operation. No scrap is charged in normal operation. The production figures for the demonstration (393 kg new pig-iron in 2 hours 27 minutes) extrapolate to 3.8 tons new pig-iron in 24 hours.

Let us now arbitrarily suppose that in normal operation production was the same as in chapter 2, i.e. *c.* 700 kg pig-iron in 24 hours. The charcoal and blast-air consumption can be scaled down proportionally and still be in balance. The blast input would therefore be $(700/3800) \times 9 = 1.6\ m^3/min$ instead of $9\ m^3/min$ in the demonstration. This is the same as the figure given in chapter 2; which should not be surprising since similar amounts of charcoal are actually consumed in each case. (The average stroke rate of the windbox would be $(700/3800) \times 61 = 11$ strokes/min.)

In the description translated in chapter 2 (section 2.2.5.2) one person works the windbox to produce this volume of blast. In the Huang Jiguang furnace the furnace burden is much less permeable, for two reasons: the smaller total amount of charcoal charged, and the use of crushed charcoal. Thus the labour involved in producing the blast would be much harder. Perhaps two persons could do the work, or perhaps it might require many more.

Thus it is not obvious which furnace is economically more efficient. The Huang Jiguang furnace is presumably more efficient in fuel consumption, but it requires more labour, perhaps much more.

CHAPTER 6

REFINING FURNACES

6.1. Other descriptions

In the description translated in chapter 2 (section 2.3) the refining furnace and its operation are described rather briefly; it will be useful to supplement it with a description of a similar furnace in southern Henan, also published in 1958.

A brochure compiled by the Jiangsu Provincial Office of Metallurgical Industry (*Lian'gang* 1958) includes six articles on local refining techniques of different parts of China. One of these articles (pp. 11-22) describes a technique used in Shangcheng County, in southern Henan. Two others (pp. 1-9, 24-9) give additional notes based on experience using the Shangcheng technique in Nanjing. The following description is based on these three articles (primarily pp. 11-16).

Generally two furnaces are built together and used alternately. Figures 27-8 show such a pair of furnaces. First a base is prepared: a rectangular pit is dug, 170 × 100 cm, depth 90 cm. This is filled with solidly rammed earth. The material used is a local yellow soil which contains 35-45% coarse sand.[1]

The two furnace cavities are cut into this rammed-earth base. The cavity is 'tangerine-shaped' (see figures 28-9), with dimensions:

maximum diameter	62 cm
bottom diameter	28 cm
mouth diameter	31 cm
depth	38 cm

The volume of this cavity is about 0.06 m^3. Its shape is very important for the uniform circulation of the blast in the cavity.

A cover is placed over the cavity, covering about three-quarters of the mouth. This cover seems to be the most difficult part of the furnace construction. It is made of a mixture of clay and sand, blown dry in a dark place for 7-10 days, then baked over a slow fire. It is about 5 cm thick. A blast-hole, about 3 cm in diameter, is cut about 8 cm from the edge.

The cover is positioned so that the blast-hole is precisely over the centre of the cavity. A tuyère and blast-pipe are fitted as shown in figure 29. It is very important that the blast should be directed precisely downward to the centre of the cavity.

The uncovered part of the mouth of the cavity is widened to form a rectangular 'furnace door', 17 × 20 cm. Because of the violence of the puddling

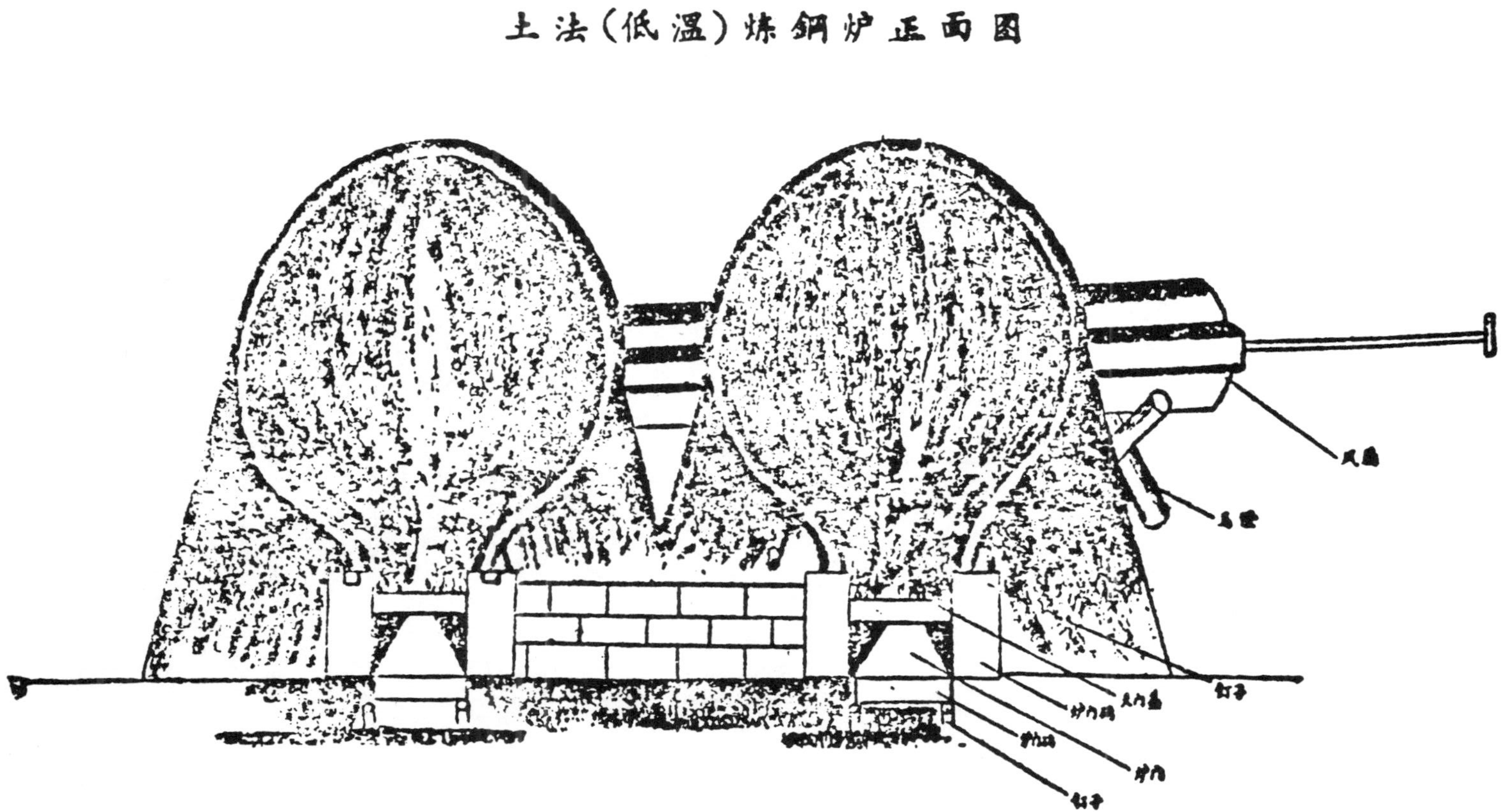

Figure 27. Pair of iron-refining furnaces in Shangcheng, Henan, 1958; manually-operated windbox in the background (*Lian'gang* 1958, p. 18).

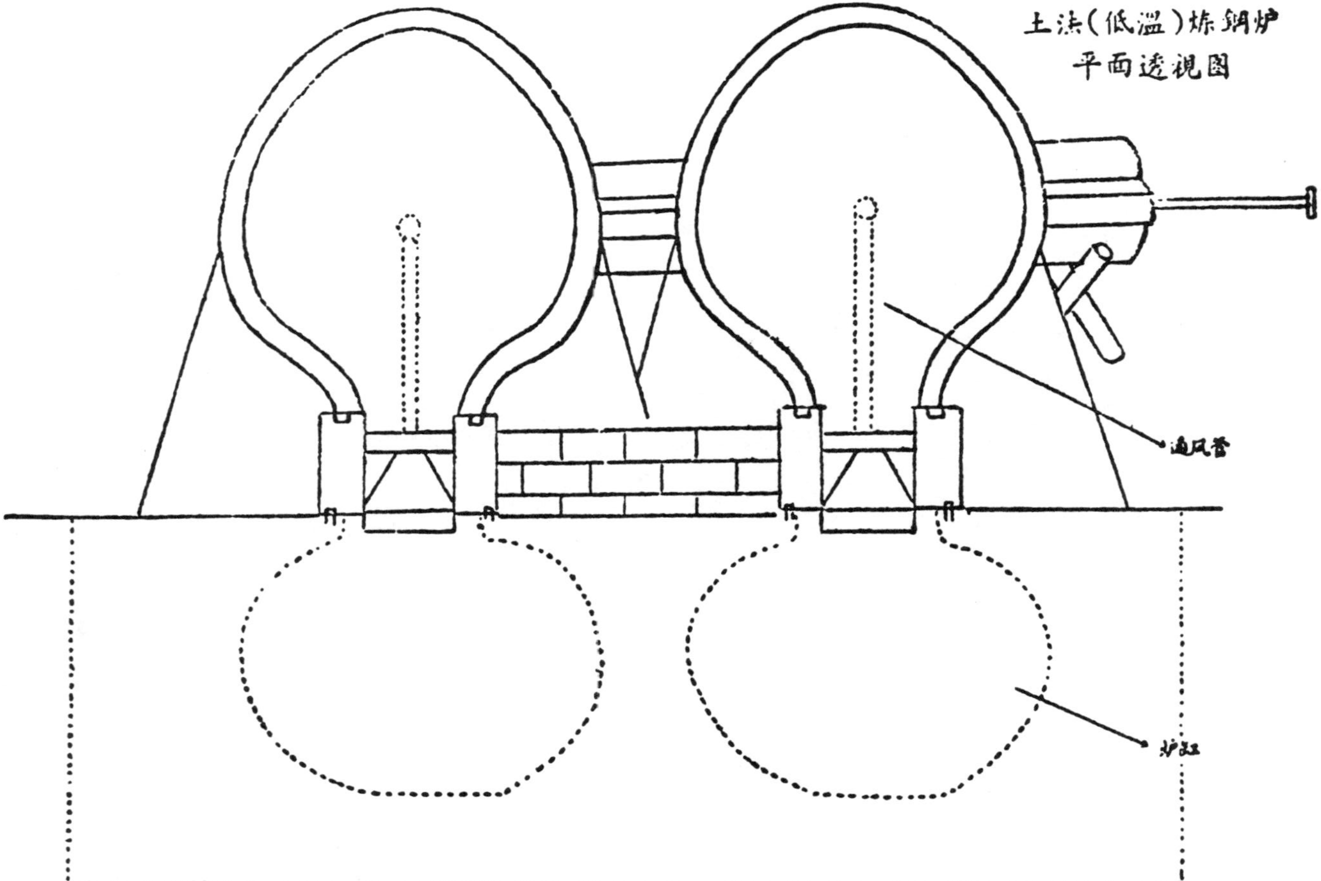

Figure 28. Diagram of a pair of iron-refining furnaces used in Shangcheng, Henan (*Lian'gang* 1958, p. 20).

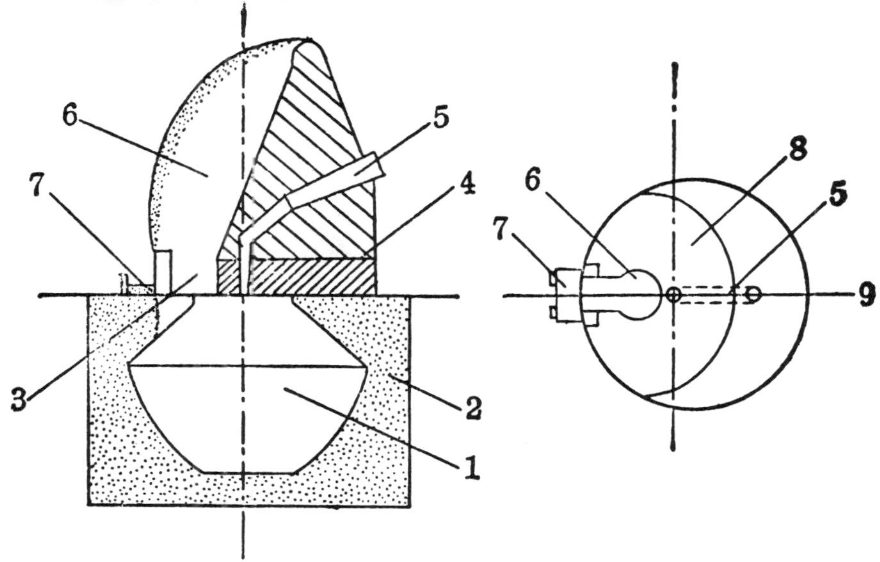

插图 60　河南省商城县土法炒钢炉的正视图和剖视图
(采自冶金工业出版社编《土铁土法炼钢》1958 年 10 月版，
第一篇《河南商城县的土法低温炼钢》)
部位名称：1.炉缸　2.夯实的耐火泥　3.炉门口　4.天门盖　5.通风管
6.炉门　7.炉门铁　8.炉窝　9.地面

Figure 29. Diagram of an iron-refining furnace used in Shangcheng, Henan, 1958 (Yang Kuan 1982, p. 225; originally in *Lian'gang* 1958, p. 23).

1. Hearth. 2. Tamped fireclay. 3. Furnace opening. 4. Cover. 5. Windpipe. 6. Furnace opening. 7. Iron reinforcements. 8. Nest. 9. Ground level.

炼钢用工具、器具图

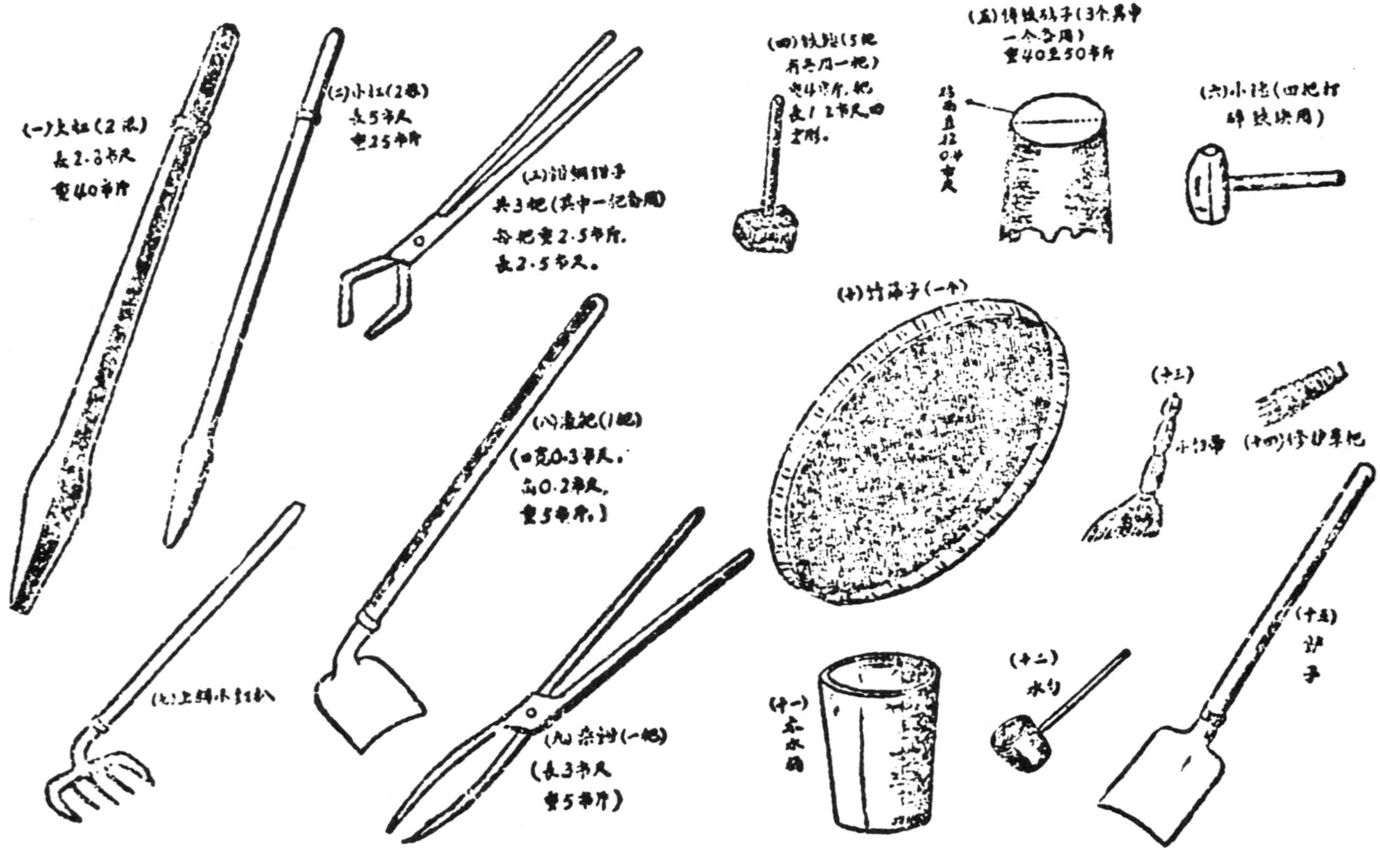
(一)大钎(2根)
长2.3市尺
重40市斤
(二)小钎(2根)
长5市尺
重25市斤
共3把(其中一把备用)
每把重2.5市斤.
长2.5市尺。
重40至50市斤
(六)小锤(四把打碎铁块用)
(口宽0.3市尺.
高0.2市尺.
重5市斤.)
(长3市尺
重5市斤)
(十二)
水勺

operation, the sides of the furnace door are reinforced with slabs of cast iron.

Over the furnace is built a 'nest' of yellow sandy clay mixed with a little sand and water. This is piled up to a height of about 70 cm and formed as a kind of surrounding wall as shown in figures 27-9. The nest apparently serves to preserve heat and to protect the surroundings (especially the wooden windbox) from fire.

The furnace cavity is dried with a slow fire, then baked hard with a hot fire under full blast. This takes 2-3 hours.

Operation is by shifts of eight workers working twelve hours per shift. The workers are: one puddler, two tong-men (who withdraw the puddled balls from the furnace with tongs and take them to the anvils), four hammer-men, and one windbox-operator. The tools used are shown in figure 30.

The refining furnace can be charged either with cold pig-iron or with molten pig-iron direct from the blast furnace. Using molten iron gives a higher yield of wrought-iron from pig-iron (*c.* 90% vs. *c.* 85%), and saves both time and fuel; on the other hand the co-ordination of the smelting and refining operations is probably not a trivial organizational problem.

When cold pig-iron is charged, there are eight or nine furnace cycles in one 12-hour shift. In each cycle *c.* 140 *jin* (70 kg) of wrought-iron is produced.[2]

The sequence of each cycle is as follows:

1. Charge 7 *jin* (3.5 kg) of wood, 100 *jin* (50 kg) of broken-up pig-iron, and 8 *jin* (4 kg) of charcoal.
2. Ignite, close the furnace door, and work the blast for 15-20 minutes.
3. When the iron is red-hot, open the furnace door and stir the charge with an iron bar.

Figure 30 (opposite). Tools and accessories used in iron-refining in Shangcheng. Henan (*Lian'gang* 1958, p. 22).

Left to right, top to bottom:

1. Large iron bar (2 used). (Length illegible), weight 40 *jin* (*c.* 20 kg). 2. Small iron bar (2 used). Length 5 *chi* (*c.* 167 cm), weight 25 *jin* (*c.* 12.5 kg). 3. Tongs for removing wrought-iron balls (2 used + 1 spare). Length 2.5 *chi* (*c.* 83 cm), weight 2.5 *jin* (*c.* 1.25 kg). 4. Iron hammer (4 used + 1 spare), for forging refined balls. Weight 4 *jin* (*c.* 2 kg), handle length 1.2 *chi* (*c.* 40 cm). 5. Anvil (2 used + 1 spare). Weight 40-50 *jin* (*c.* 20-25 kg), diameter of upper face 0.4 *chi* (*c.* 13 cm). 6. Small hammer (4 used), for breaking up pig iron. 7. Small rake used in charging. 8. Hoe (1 used). Breadth of head 0.3 *chi* (*c.* 10 cm), length of head 0.2 *chi* (*c.* 7 cm); weight 5 *jin* (*c.* 2.5 kg). 9. Firewood tongs (1 used). Length 3 *chi* (*c.* 1 m), weight 5 *jin* (*c.* 2.5 kg). 10. Bamboo sieve (1 used). 11. Water bucket. 12. Water dipper. 13. Small broom. 14. Straw brush for cleaning furnace cavity. 15. Shovel.

4. Charge an additional 5 *jin* (2.5 kg) of wood, 66 *jin* (33 kg) of broken-up pig-iron, and 6 *jin* (3 kg) of charcoal.
5. Close the furnace door and continue working the blast.
6. After about 10 minutes white flames appear around the furnace door, the furnace cavity is white hot, and the surroundings have a dull red heat.

At this point puddling can begin. The following is a translation of the instructions for puddling. The obscurities in the translation reflect obscurities in the original; this operation is obviously better learned by experience than described in words.

Open the furnace door and stir up the charge with the large iron bar. When the pig-iron begins to fragment, puddle where it is fragmenting. With the iron bar move the red-hot small pieces of iron under the blast, so that they fragment. Then move the surrounding unfragmented pieces of iron to the centre for refining. Continue in this way; after about 15 minutes most of the iron will be in a pasty state. With the small iron rod gather the pieces directly under the blast into a ball of 2.5-3 *jin* (1.25-1.5 kg), about the size of two fists. Expose all sides of it to the blast until it is white hot, then remove it to the anvil with the tongs.

If, because of incorrect operation, the iron aggregates into a large clump, the clump should be moved under the blast. The blast is reduced for a moment, and the clump struck forcefully a couple of times at the point where the blast hits it. This should fragment the clump and prevent its growing larger and larger.

When the steel ball is taken to the anvil it should be hammered lightly at first, to avoid breaking it up. It is formed into a rectangle. After it has cooled to purplish-red it may be hammered with force. After it is black it should be hammered lightly a few more times. In general each ball is struck a hundred or more times in all.

(*Liang'gang* 1958, p. 15)

The following analysis is given for one sample of wrought-iron made by this method:

C	Si	Mn	S	P
0.17%	0.21%	0.04%	0.047%	0.17%

When molten iron is charged directly, operation is as follows. The furnace is charged with 7-10 *jin* (3.5-5 kg) of charcoal. Then 200 *jin* (100 kg) of molten iron is poured in. The molten iron should be rather viscous; if it is too hot the furnace walls can be damaged. The blast is started, and puddling begins immediately. The puddling operation is said to be the same as above. After 15-20 minutes the wrought-iron balls can be withdrawn and hammered. One furnace cycle takes about one hour.

The refining process in southern Henan was also described by Nyström in 1917. This description is quoted briefly by Tegengren (1924, pp. 335-6):

80 catties (48 kg) of broken up pig-iron is put together with wood into each furnace, and a strong wind, striking the charge from above, is kept up. The mass is stirred by a rod operated through the furnace door. At last a spongy lump of wrought iron remains and this is picked out with a pair of tongs and rapidly pounded on an anvil until a short bar, 3 × 3 × 11 cm, is obtained.

An analysis of the iron bar shows:

Fe	C	Si	P	S	[total]
98.68%	0.34%	0.37%	0.16%	0.045%	[99.595%]

This iron is carried by coolies and marketed all over South Henan.

6.2. Variants of this refining furnace in other parts of China

We have above considered three descriptions of refining techniques practised in southern Henan, two from 1958 and one from 1917. Despite variations in detail it is clear that all three describe essentially the same furnace and operating procedures. Variants of the technique seem to have been used throughout China in the early twentieth century, as can be seen for example in Tegengren's survey.[3]

As usual throughout the present study my primary concern here is the traditional techniques of one limited region, southern Henan; techniques of other parts of China must be treated more briefly.

The most interesting of the variants are those in which the fuel and the iron are separated, so that coal (if it has relatively low sulphur) can be used instead of charcoal. The basic principle here is that the fuel is burned in one chamber, into which the blast is blown; the flame from this chamber is led to a separate refining hearth, where puddling takes place. Yang Kuan (1960, pp. 184-91) describes several different implementations of this principle, used in different parts of China; some of these are shown in figures 31-2. He refers to these as 'reverberatory furnaces' (*fanshe lu*), but this term would seem to be inaccurate: the iron is heated only by the flame, not to any great extent by reflected heat.

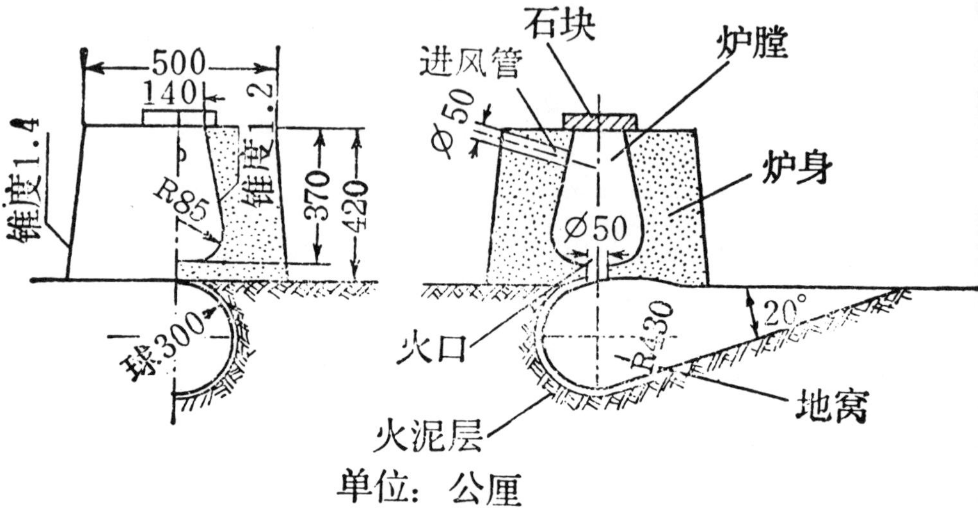

插图 62 西安的一种简单反射炉
（采自科技卫生出版社编《土法低温炼钢》第六篇
《最简单的反射炉炼钢》，1958 年 10 月出版）

Figure 31. Refining furnace used in Xi'an, Shaanxi, in 1958 (Yang Kuan 1982, p. 228). Dimensions are given in mm. Coal or charcoal is burned in the upper chamber. Blast is blown in through the slanted passage near the top; the flame blows down into the lower chamber, where the iron is puddled.

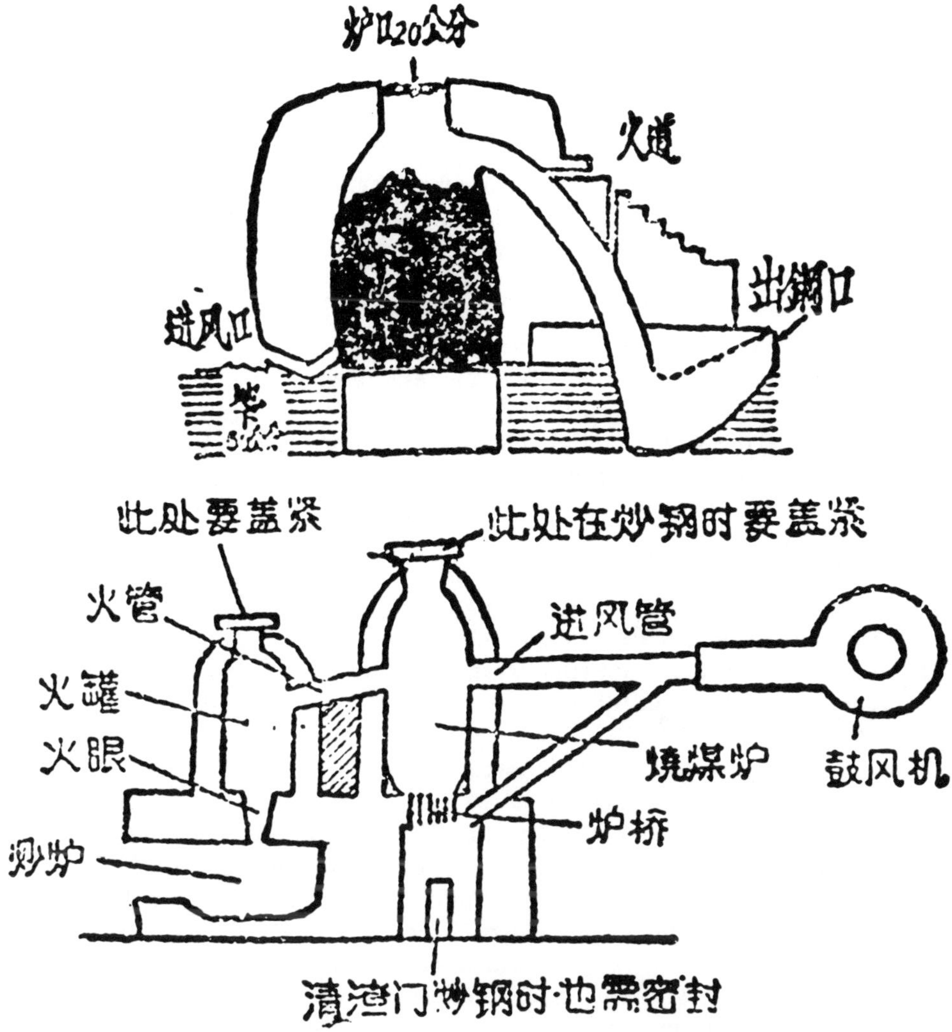

Figure 32. Two types of refining furnace used in Sichuan in 1958 (Yang Kuan 1960, p. 187). Dimensions not given.

Above: Coal or charcoal is burned in the large chamber; blast is blown in at the left. The flame travels through the long passage to the chamber at the right, where the iron is puddled.

Below: Anthracite coal is burned in the firebox at the upper right. Ashes fall into the ash-box at the lower right. Blast is blown into both the firebox and the ash-box. The flame travels through the large empty chamber and then into the chamber at the lower left, where the iron is puddled.

6.3. *Iron-refining in ancient China*

Chinese archaeologists have noted several artifacts of the first and second century A.D. made of wrought-iron or steel which appears to have been produced by refining pig-iron rather than by the older bloomery process. The two processes can be distinguished by microscopic examination of slag inclusions in the iron.

Figures 34 and 36 show slag inclusions in some artifacts of bloomery iron. These can be compared with figures 37 and 39, from artifacts of the Eastern Han period (A.D. 25-220); the latter artifacts are considered to be of refined pig-iron because of the small total amount of slag, the small size of the inclusions, and their uniform distribution.

In addition, the knife in figure 37 has a uniform carbon content of 0.6-0.7% throughout. Much more variation in carbon content would be expected if this steel had been produced by the cementation of bloomery iron (Li Zhong 1975, p. 15).

This metallographic evidence seems very convincing, but it is essential that it be confirmed by comparison with modern materials of known provenance. Since the traditional refining technique has been in use as recently as 1958, it should be relatively easy for Chinese archaeologists to obtain objects made by this method and compare their microstructures with those found in ancient artifacts. Gordon (1983) gives microphotographs of slag inclusions in nineteenth-century American bloomery iron and fined iron, and these tend to confirm the Chinese archaeologists' view.

I have investigated the microstructure of a kitchen knife which was made in Guangzhou (*Shuangshi,* 'two lions', brand), probably around 1980 (purchased in San Francisco in 1982). It contains numerous uniformly distributed inclusions which resemble those seen in figures 37 and 39, so it seems likely that the ancient refining process is to some extent still in use today.

The question of when the finery hearth came into use in Europe would seem to be highly important in the history of technology: it is therefore surprising that almost no work has been done on the subject from a metallographic point of view. Johannsen (1953, p. 148) states, 'The higher FeO content of slag inclusions [in fined iron] makes it possible, according to research by Karl Daeves, to determine whether ancient wrought-iron was produced by the direct or the indirect process.' Unfortunately he does not give a precise reference to Karl Daeves's publication, and I have been unable to find it.

Even earlier than this metallographic evidence are the remains of several actual refining furnaces which are rather similar to the one in the translated description. These have been found in excavations of three ironworks sites of

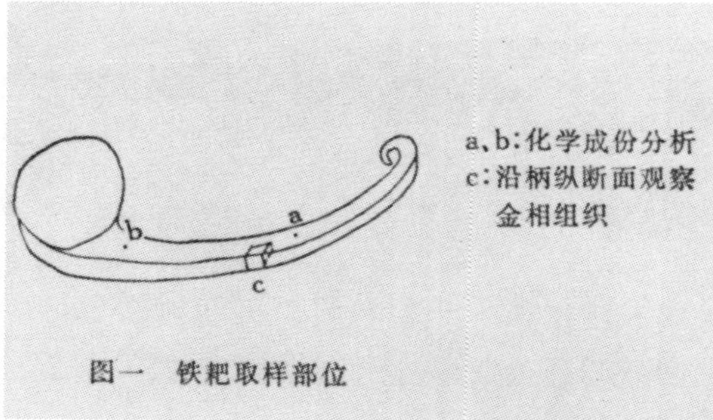

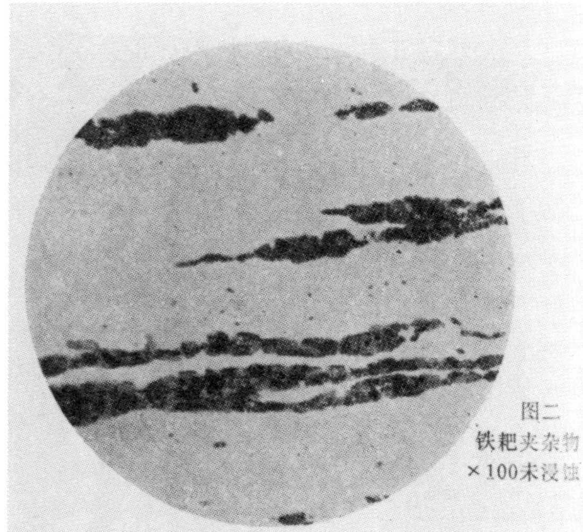

Figure 33 (left). Wrought-iron gouge from an ancient copper-mine site at Tonglüshan in Daye County, Hubei. Length of handle 50 cm. The sample for metallographic examination was taken at *c*. This copper-mine was worked at least from the 5th to the 1st century B.C., probably both earlier and later (Ye Jun 1975, pp. 19, 24; WW 1975.2:1-12; Buck 1975; Hsia Nai 1975).

Figure 34 (right). Micrograph, 100×, unetched, showing slag inclusions in the gouge of figure 33 (Ye Jun 1975, p. 24).

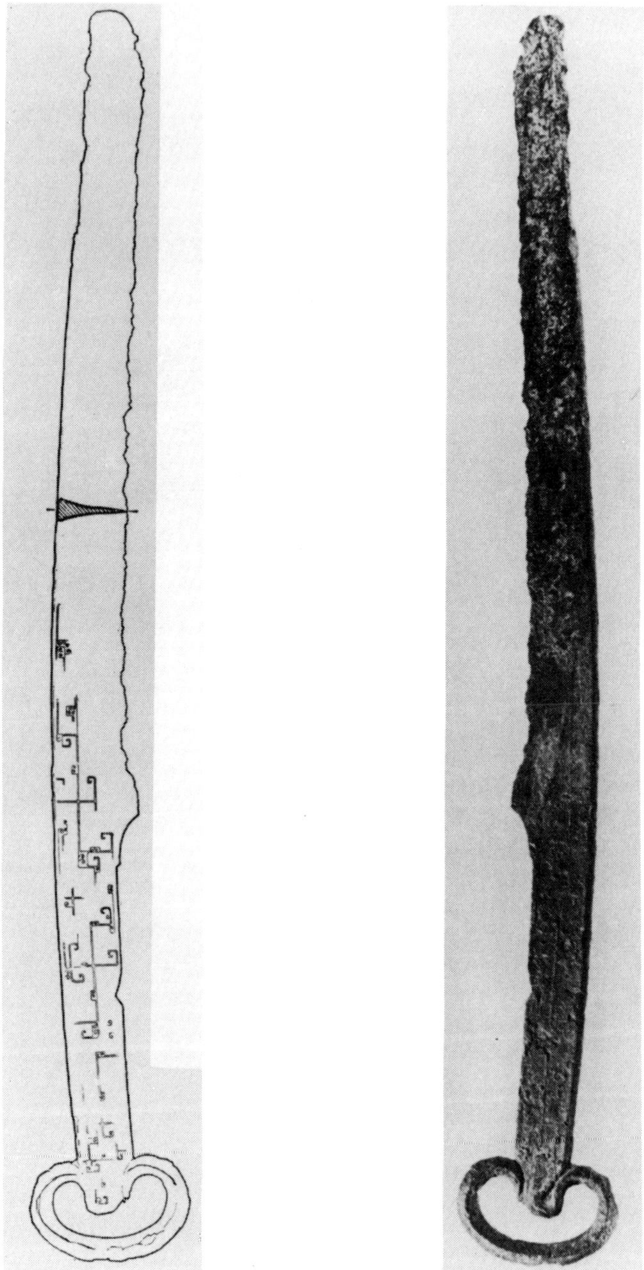

Figure 35. Gold-inlaid knife from Tomb no. 1 in Mancheng, Hebei (artifact no. 1:5197). Total length 42.4 cm (*Mancheng* 1980, pp. 105-7 and plate 68.1).

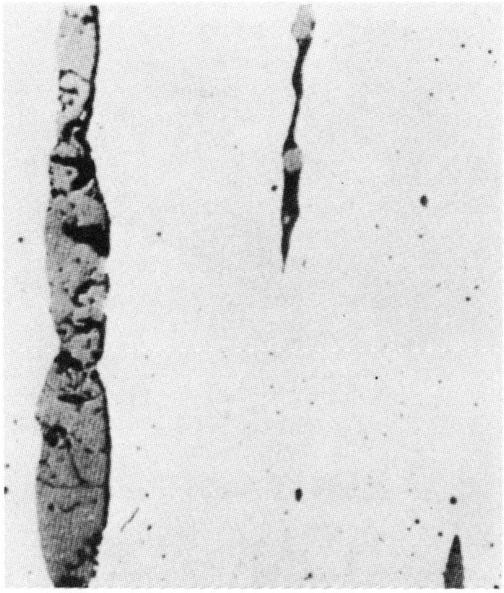

Figure 36. Micrograph, 180×, (unetched?), showing slag inclusions in a section of the knife of figure 35. Black areas: FeO–2FeO·SiO_2 eutectic; grey areas: FeO. The larger inclusion at the left was probably introduced by the smith in forge-welding two pieces of iron together. The smaller inclusions to the right, which contain much more silica, are probably bloomery slag. (*Mancheng* 1980, pp. 372-3 and plate 259; see also Li Zhong 1975, p. 12 and plate 5.25).

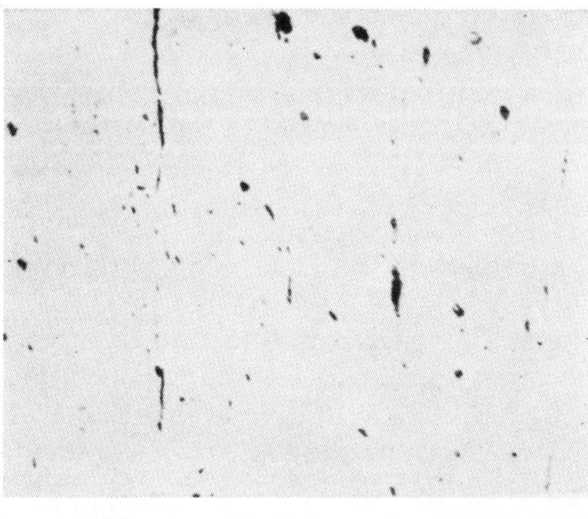

Figure 37. Micrograph, 200×, (unetched?), showing slag inclusions in a steel knife from a tomb excavated at Cangshan in Linyi County, Shandong. An inscription on this knife gives a date equivalent to A.D. 112 (Liu Xinjian & Chen Zijing 1974; Li Zhong 1975, p. 15 and plate 6.35).

Figure 38. Kitchen knife, dated Eastern Han (A.D. 25-220), found near Nanyang, Henan. Dimensions 11.2 × 17 × 0.5 cm (KGXB 1978.1:22-23 and plate 2.5).

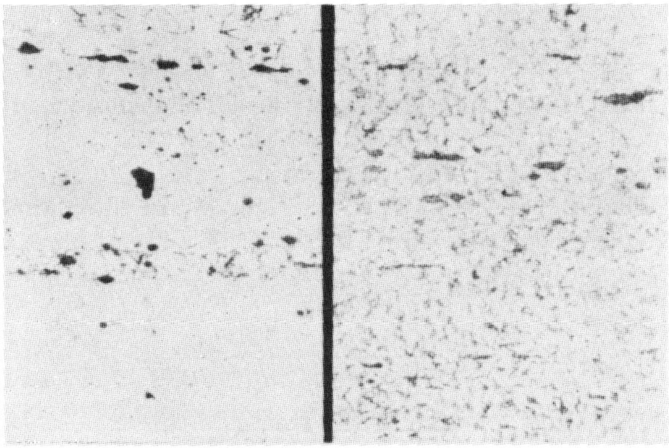

Figure 39. Two micrographs, 160×, (etched with nital?), showing slag inclusions in the kitchen knife of figure 38 (KGXB 1978.1:22-23 and plate 2.5).

the Han period (206 B.C.-A.D. 220), all in Henan province: in Gongxian, Fangcheng, and Nanyang.

The ironworks site at Tieshenggou in Gongxian County, from the evidence of artifact types and coins, appears to have started operation during the reign of Emperor Wu (140-87 B.C.) and to have been abandoned during the reign of Wang Mang (A.D. 8-24) (*Gongxian* 1962, p. 38). One refining furnace was found, shown in figures 40-41.

The ironworks site at Wafangzhuang, near Nanyang, was excavated in 1959 and 1960, but a full excavation report has never been published. It appears to have been in operation from the middle of the Western Han to the late Eastern Han: roughly first century B.C.-second century A.D. (KGXB

2. 低溫炒鋼炉(炉 17)

Figure 40. Remains of a refining furnace excavated near the Han-period ironworks site at Tieshenggou in Gongxian County, Henan (*Gongxian* 1962, plate 6.2).

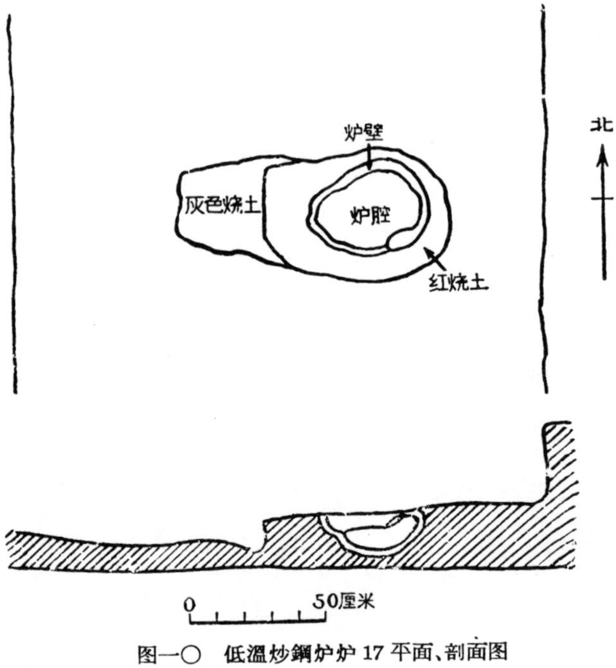

Figure 41. Diagram of the Han-period refining furnace shown in figure 40 (*Gongxian* 1962, p. 13). The scale shows cm.

1978.1:3). 'Several' refining furnaces were found on the site; one is diagrammed in figure 42.

The ironworks site at Zhaohecun in Fangcheng County was excavated in 1958 and 1976, but no excavation report was published. It has been tentatively dated to the Han period (KGXB 1978.1:3). Six refining furnaces similar to the one in Gongxian were found here; no details are available (*Gongxian* 1962, p. 11).

From figures 40-42 it can be seen that these furnaces are rather similar to those described above for the twentieth century. A shallow cavity was dug and lined with fireclay. Whatever superstructure there may have been is now lost. The shape of the cavity is rather more open than in the later furnaces.

It may be that we even have a picture of iron-refining in the Han period. Figure 43 shows a detail of a tomb-relief unearthed in 1930 at Hongdaoyuan, in Tengxian County, Shandong (WW 1959.1:2; for the full context of this detail see *Corpus* 1950, vol 1, plate 73). It can be dated by stylistic criteria to the first or second century A.D. It seems clear that the relief depicts metal-

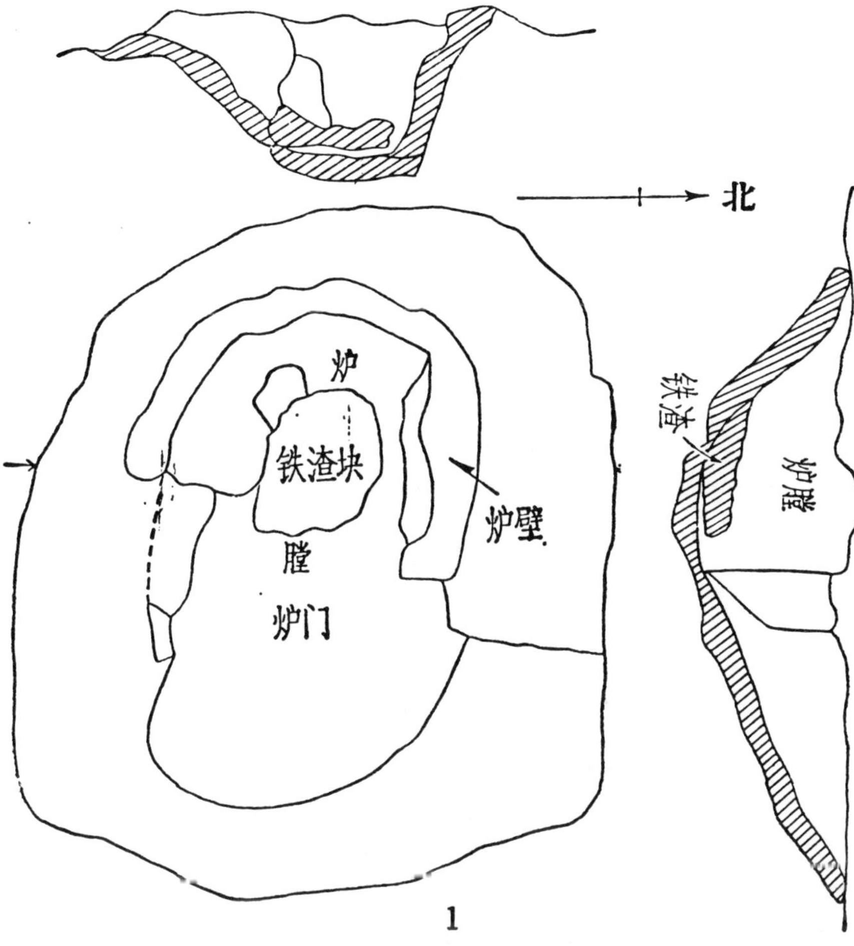

图八 瓦房庄19号炒钢炉平、剖面

Figure 42. Diagram of the remains of a refining furnace at a Han-period ironworks site at Wafangzhuang, Nanyang, Henan (KGXB 1978.1:22). Dimensions not given; presumably about the same as in figure 41.

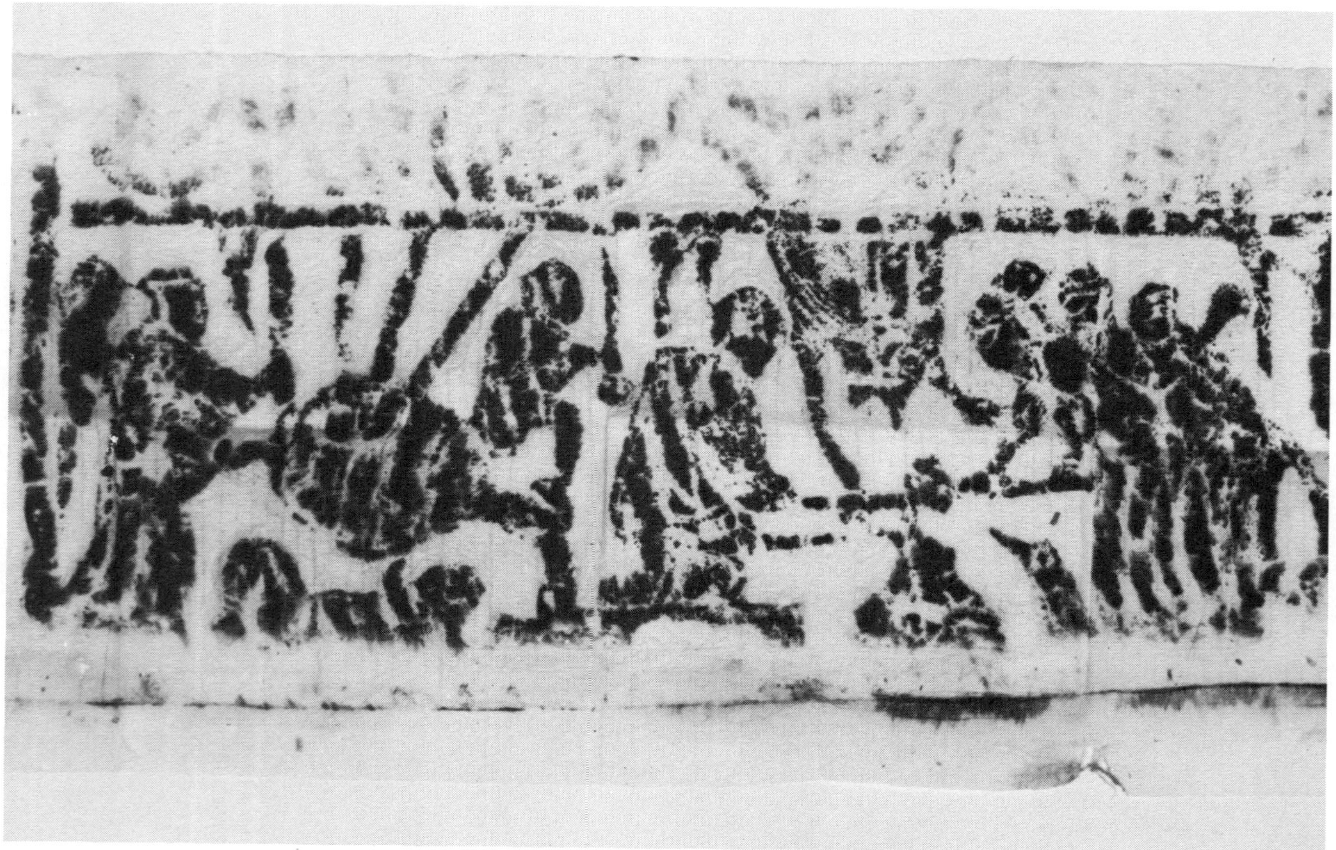

Figure 43. Detail of a rubbing of an Eastern Han tomb-relief discovered in 1930 at Hongdaoyuan, Tengxian County, Shandong. Archives of the East Asian History of Science Library (Needham Research Institute), Cambridge, England.

work of some sort, but there has been considerable discussion as to how it should be interpreted. One suggestion is that the detail in figure 43 shows iron-refining using a furnace like the one we have been discussing (KGXB 1978.1:22). The bellows has been reconstructed as in figure 44. The blast is directed downward, presumably into a cavity dug into the ground. The worker standing at the left works the bellows. The activities of the next two workers, one lying down and one standing, are unclear. The next may be holding a piece of hot iron with tongs on an anvil while several (perhaps four) workers hammer it.

For the time after the Han and before the twentieth century the available information on iron-refining is rather sparse.

Metallographic examination of wrought-iron and steel artifacts from a shipwreck of the fourteenth century A.D. indicate, in the same way as discussed above, that they were produced by refining pig-iron (KG 1978.6:400-1). This is the only relevant archaeological material published so far.

There are many early written sources which mention iron-refining, but I have not yet studied these. Needham does not go into the subject in detail, but he quotes, in various connections, several sources which mention it (1958, pp. 15, 16, 28, 29, 33, 34, 36, 37). A few more sources are quoted by Li Zhong (1975, p. 15) and by Yang Kuan (1960, pp. 174-6).

6.4. Comparison with early Western refining techniques

In any iron-refining process the pig-iron is heated to a high temperature and exposed to air and an oxidizing slag (rich in FeO). This oxidizes the carbon in the iron, together with other impurities (silicon, sulphur, manganese, etc.), and a small amount of the iron itself. The reactions involved are exothermic, so that the temperature tends to rise as the decarburization proceeds. Figure 45 shows the iron-carbon equilibrium diagram with some very rough approximations (not much more than guesses) of the variation of temperature and carbon content through time for three Western pre-modern refining processes: the old 'fining' process, puddling, and the Bessemer process.

Percy (1864, pp. 579-692) describes a great many versions of the fining process used in different places, as well as the puddling process. The following generalized description is based largely on his descriptions (see also den Ouden 1981-2).

In the various fining processes pieces of pig-iron are mixed with a large amount of charcoal and exposed to a strong blast of air. The operator uses an iron bar to stir up the charge, ensuring that all parts are exposed to the blast and breaking up any aggregates which form. Through most of the process

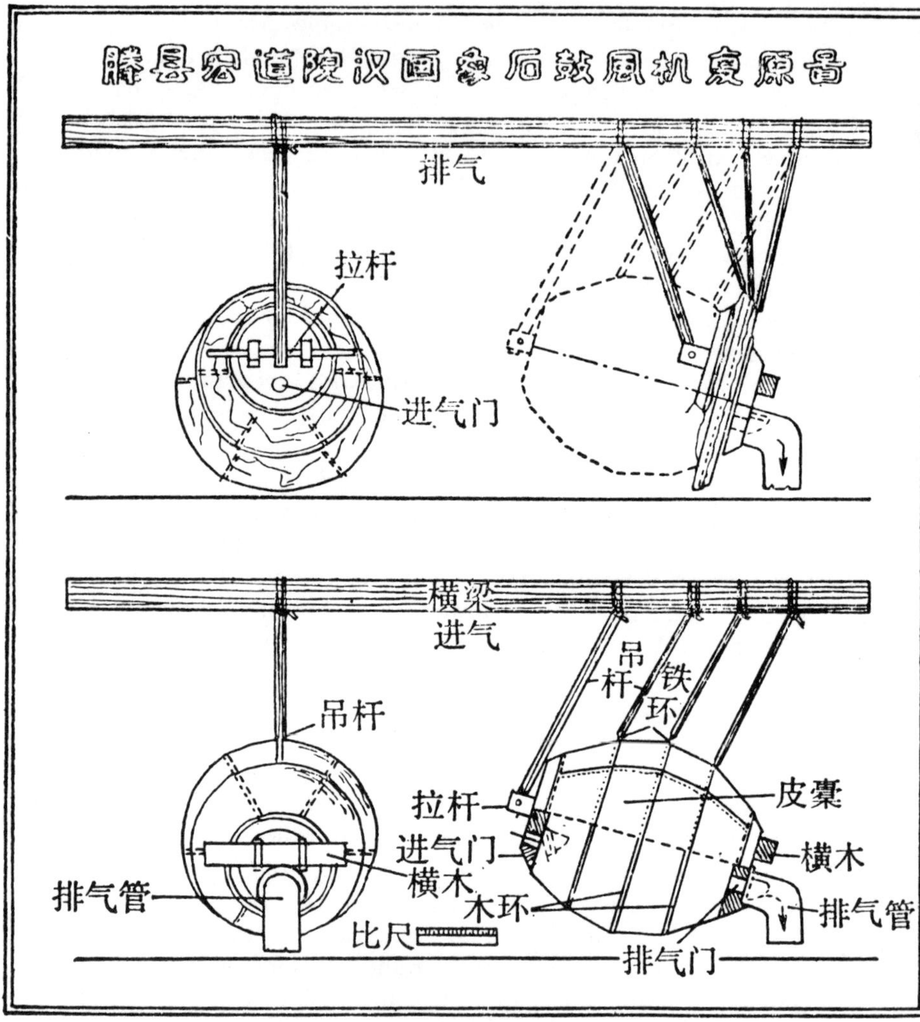

插图 11 滕县宏道院汉画像石鼓风机复原图
(采自«文物»1959 年第 5 期王振铎«汉代冶铁鼓风机的复原»)

Figure 44. Reconstruction of the bellows shown on the left in figure 43 (Yang Kuan 1982, p. 91; originally in Wang Zhenduo 1959, p. 43).

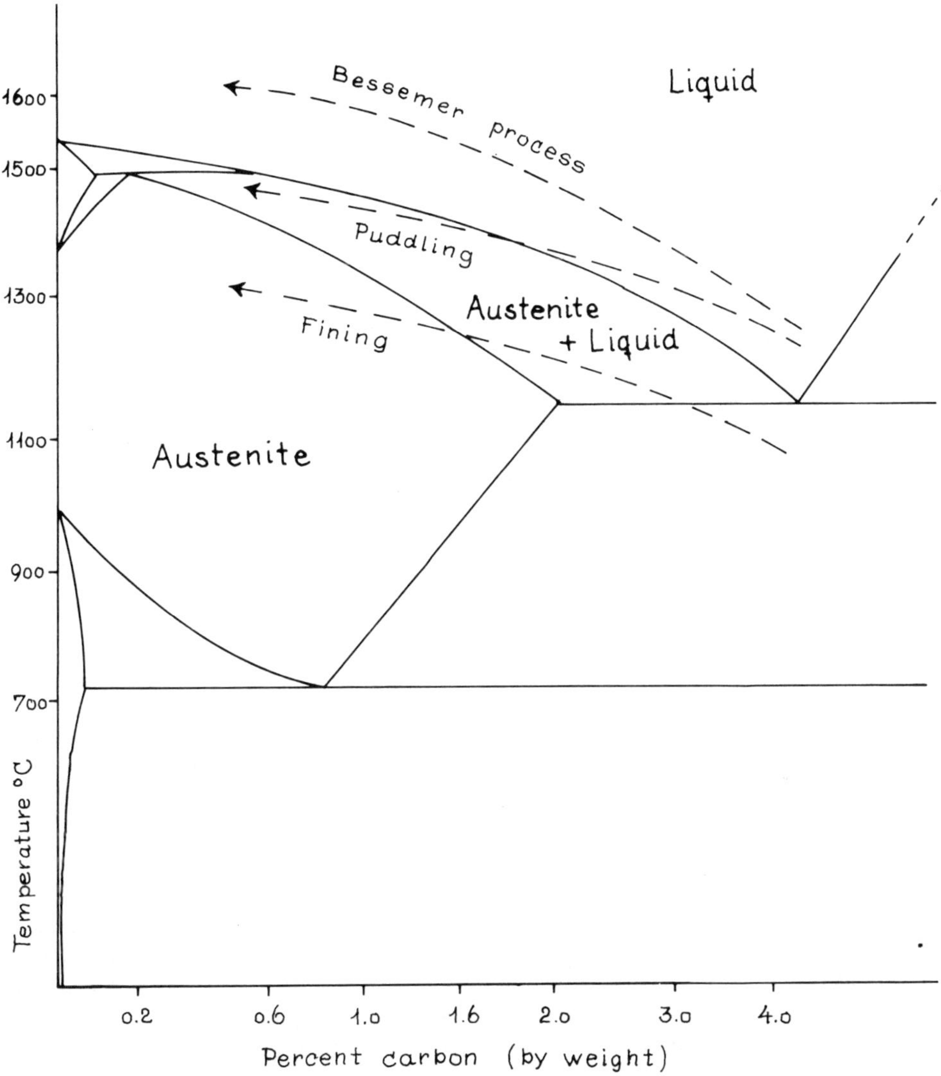

Figure 45. The iron-carbon equilibrium diagram. Broken arrows give very rough approximations for the variation of temperature and carbon content through time in three pre-modern iron-refining processes.

none of the iron is in a liquid state. The iron must be kept broken up into small fragments because diffusion of carbon to the surface of the solid iron is an important part of the process. The large amount of charcoal mixed with the iron functions not only to provide heat but also to keep the iron fragments physically separated.

In this process it was believed to be essential that the temperature be kept fairly low. This necessity was probably related to the removal of phosphorous: the oxidation of phosphorous in iron proceeds to a significant extent only at lower temperatures.[4]

The puddling process, which was patented by Henry Cort in 1784 and soon largely replaced the earlier processes, is different in two significant ways: (1) the operation takes place in a reverberatory furnace, with the iron and the fuel separated; (2) higher temperatures are used, so that the iron is molten through most of the process. (For more on the puddling process see Mott 1983.) Probably the first point was the most important, since it allowed the substitution of coke for charcoal in iron-refining. The second point probably followed naturally from the first: it was impracticable to keep solid fragments of iron from aggregating together in the reverberatory furnace. On the other hand, in the fining hearth it was impracticable to keep iron molten, and keep it in an oxidizing environment, when the heat was provided by fuel in the same hearth.

In the puddling process molten pig-iron, heated by flame and reflected heat from a coke fire, is stirred about with an iron rod in order to keep it exposed to air and the oxidizing slag throughout. Oxygen dissolves in the liquid metal and oxidizes the carbon and other impurities in it. For most of the duration of the operation the temperature is kept above the melting point (liquidus) of the iron; when the carbon content has fallen so low that the temperature is below the liquidus line (see figure 45), the iron begins to 'come to nature': iron with a fairly low carbon content precipitates out of the melt.

In the puddling process the removal of phosphorous appears to take place not by oxidation but by concentration in the remaining liquid phase as the iron comes to nature. The great problem in the early years of the Bessemer process was that it did not remove phosphorous: at the high temperatures involved, phosphorous was not oxidized; and since liquid steel was produced there was no chance of using differential solubility to remove it.

In the Chinese refining furnaces of southern Henan described earlier the mixture of iron fragments and charcoal in the furnace cavity is stirred up by the operator in essentially the same way as in the early Western fining processes. The Chinese word for this operation, used at least since the eleventh century A.D. (Needham 1958, p. 34), is *chao*. The original meaning of this word is 'stir-frying', one of the basic processes in Chinese cooking. The English word 'puddling' is normally translated by *chao* in modern Chinese, and I have followed this convention in translating *chao* as 'puddling'. Nevertheless it does seem that the process described is more closely related to the earlier fining processes than to puddling: the iron and fuel are mixed together, and the iron is never molten.

On the other hand the variants of the process described in section 6.2 above do seem to deserve to be called puddling. The iron and fuel are separated, and this seems to be the basic distinctive feature of puddling. I have no information on whether the iron is decarburized in the liquid or solid state. It seems likely that it is in the liquid state, because of the practical difficulty, mentioned earlier, of keeping small fragments of solid iron from aggregating together in a hearth with no fuel.

A point that should be noticed here is the remarkable fuel economy of the refining processes of southern Henan. In any refining process, though the reactions involved are exothermic, some fuel is required to reach the necessary temperature and to balance heat losses. In the processes of southern Henan, for each ton of wrought-iron produced, about 120-160 kg of charcoal and 120-140 kg of wood are used. Percy (1864, pp. 601, 607, 610) gives fuel-consumption data for several fining processes: they vary from 900 to 3000 kg of charcoal per ton of wrought-iron.

This difference can no doubt be attributed to the great attention paid to heat-insulation in the construction of the Chinese refining furnaces. The early Western fining hearths seem to have been large, rather open structures, from which a great deal of heat must have been lost. This waste of heat may have been necessary in order to maintain control over the temperature and keep it low enough to ensure removal of phosphorous.

The ancient Chinese furnaces discussed in section 6.3 above seem to have been much more open than the later ones. Probably heat-loss was much higher in the ancient furnaces, and fuel-consumption correspondingly high.

The labour economy of the process also seems rather good. At the beginning of this chapter we have seen that eight workers produce about 600 kg of wrought-iron in 12 hours. Of these workers one operates the blast, four hammer the iron, and only three work at the furnace. When these figures are compared with those of the early West it must be remembered that the latter used water-powered blowing engines and hammers. In China water-power has been used for both operations since very early times (Needham 1965, pp. 51-2, 224-5, 369-80, 390-96); the reasons for its not being used in southern Henan in 1958 presumably lie in local geographic and economic conditions: availability of suitable water-courses, availability of investment capital, and wage levels. The labour required *at the furnace,* i.e. excluding operations which could have been water-powered, would be about 60 worker-hours per ton of wrought-iron.

Percy (1864, pp. 602-8) gives quantitive labour data for only one of the fining processes he describes: this is the Franche-Comté process, considered to be 'the most economical of all'. Here three workers work a shift of eight hours in which four balls are produced, each 80-85 kg (after shingling and before forging). It may therefore be calculated that the labour required to produce one ton of wrought-iron is about 70 worker-hours. Puddling, again as described by Percy (1864, pp. 654-8), seems to be much more labour-efficient: apparently about 16 worker-hours per ton, though Percy's data are a bit difficult to interpret.

NOTES

Notes to the preface

1 There is one technical phase for which I have found no information: foundrywork. In fact there seems to be very little information available on traditional foundries anywhere in China. A major question is how grey-cast ploughshare-points, mouldboards, and woks were cast using the low-silicon white pig-iron from the blast furnace. On the casting of woks, a specialized branch of foundrywork in China, see CN 1884; Irresberger 1916; and Hommel 1937, pp. 28-31. Massive ceramic moulds are used; these are sometimes pre-heated to a bright red or almost white heat. McCaskey (1903a, pp. 53-8) gives some information on the casting of ploughshare-points in the Philippines, and it is possible that this will be relevant to the study of traditional Chinese methods.

2 The best discussion of the campaign that I have seen is NED 1959, pp. 12-23. See also *Yuejin* 1958; *Ganze Volk* 1959; Nikolayev & Molodtsova 1960; Yang Kuan 1960, pp. 137-54; Alley 1961a; 1961b; Hu Zhaoliang 1981, pp. 304-5. I have not yet had a chance to read MacFarquhar's (1983) study of the Great Leap Forward.

Notes to chapter 1

1 Wales 1952, pp. 152-8. The name E-Yu-Wan comes from the literary names for the three provinces Hubei, Hunan, and Anhui.

2 Nyström seems to have been an interesting person. He arrived in China at the age of 22 in 1902, and seems to have stayed almost continuously for over fifty years. He published a book and numerous articles on the geology of Shanxi province, and several popular books about China in Swedish. According to his own account he was the prime mover in the establishment of the Geological Survey of China. In 1953 he was a professor at Beijing University, and in 1956 at the University of Uppsala (Nyström 1953, 1956).

3 Zhang Youxian & Guo Yujing 1932, pp. 239-42. This article is amazing: it is obviously by working engineers with a modern education, but it is written in the purest literary Chinese and printed without punctuation.

4 At the end of 1958 the U.S. price of mild steel was $77 per ton for forging ingots and $113.50 per ton for merchant bars (*Iron age* [Philadelphia], 7 August 1958, p. 116; 1 January 1959, p. 278). Thus a ton of iron cost less than a week's salary for a U.S. steelworker. However the greater part of these U.S. prices was composed of capital costs, which were low in the U.S. and high in China, rather than labour. Therefore the above comparison may be less realistic than that arrived at by considering that $77, at the unofficial exchange rate in 1958, amounted to 63 yuan, which was about five weeks' salary for a Chinese industrial worker (see chapter 2, n. 6). (Neither comparison is very realistic.)

5 The second Plenum of the Eighth National Congress of the Communist Party, 1958.

6 It is interesting to note that the idea of developing the traditional Chinese iron industry was not new in 1958. The same theme can be seen in much earlier times; one example is the report of Guo Yujing in 1932, quoted above. An even earlier

example is a plan in 1919 at the Hanyang Iron and Steel Works to include cheap 'native' white pig-iron (0.22% S, 0.04% Si) as part of the charge in a modern blast furnace (the only one in China at the time) in order to convert it into foundry-grade pig-iron. This plan was economically feasible because of extremely high prices in China; due to the U.S. embargo on pig-iron export during the First World War the price of foundry pig in China had quintupled (C.T. Huang 1919).

Many proposals for the use of traditional Chinese iron and steel techniques were made again during the Anti-Japanese War, in the Guomindang-controlled 'Free China' region of western China. This region was almost totally cut off from the outside world, and there was no chance of importing iron. It was necessary to develop the traditional industry, which once had been very strong but now was moribund because of previous competition with cheap foreign iron. Many such proposals were published in *Kuangye banyue kan* (Mining and metallurgy semi-monthly, published in Chongqing). Needham (1948) gives a first-hand account of western China in those difficult years, with the emphasis on scientific and engineering activities.

Most of the engineers involved in the development of the traditional iron industry then were probably still active in 1958, and their experience may have influenced the decision to try again. But the traditional ironworks of western China were on a much larger scale than in the east, and they were probably much more efficient. Experience gained in one part of China was not directly applicable in other parts.

Notes to chapter 2

1 Specialized bibliographies for the campaign were included in almost every issue of the monthly *Quanguo xin shu mu* (Chinese national bibliography) in the years 1958-60.

2 According to more recent Chinese studies, the earliest archaeological evidence of the use of iron in China dates to about 500 B.C., and possible references in historical sources to iron at earlier times appear to be too uncertain to be used as evidence (Huang Zhanyue 1976).

3 The traditional Chinese 'windbox' (*fengxiang*), or double-acting piston bellows, uses an ingenious system of valves to provide blast on both cycles of the piston. It has been described many times in English, for example by Chambers (1757, pp. 13-14 & plate 18), Ewbank (1849, pp. 247-9), Hommel (1937, pp. 18-20, 28), and Needham (1965, pp. 136ff.)

4 I use the word 'refining' (translating the Chinese *lian*) in its broad modern sense, referring to any process whose purpose is to produce wrought-iron or steel from pig-iron. There is no connection with the old Western 'refinery' or 'running-out fire', whose primary purpose was to eliminate silicon from the pig-iron.

5 The basic meaning of the word *chao,* here translated 'puddling', is 'stir-frying', one of the most important processes in Chinese cooking. On the appropriateness of the word 'puddling' in this context see chapter 6, section 6.4.

6 At the official exchange rate in 1958, one yuan was equal to HK $2.34, or about £0.15 (US $0.41). On the free market the rate was about half the official rate (*New York Times,* 24 May 1958, pp. 9, 34). The average salary of an industrial worker in

China in 1958 seems to have been on the order of 50 yuan per month (Perkins 1977, pp. 40-41, 48-9). By any of these measures the initial investment of 'several ten-thousands of yuan' seems extremely high; I would have guessed, from the description given here, that the initial investment was no more than a few hundred yuan. I suspect a typographical error or some sort of misunderstanding.

7 There are several references in this text to a 'draft furnace', *choulu*. I think this is probably a term for a cupola furnace, and have translated it thus; but it is not in any of the technical dictionaries I have consulted.

8 The woods mentioned in the text are: *malishu* (Tectona grandis, 'common teak'), *jieshu* (Quercus dentata, 'daimyo oak'), and *lishu* (Q. acutissima, 'sawtooth oak').

9 See n. 4.

10 *Wok* (the Cantonese pronunciation of Mandarin *guo*) is the most common English word for the special round-bottomed braising-pan used in Chinese cooking. The worker in figure 5 is holding one of these. Most of the woks made today are of spun steel, but the remarkable cast-iron wok, with diameter 40-60 cm or more and wall-thickness as little as 1 mm, is still being produced in China. I was able to purchase one new in San Francisco in 1982. The wok is so ubiquitous in China that it is put to many uses for which it was never intended; for example it will be seen below (section 2.2.2) that the base of the blast furnace is made of three woks. More information on cast-iron woks will be found in Hommel 1937, pp. 28-34; Irresberger 1916; and Johannsen 1916.

11 The values for Fe_2O_3 and Fe_3O_4 ($FeO \cdot Fe_2O_3$) have been calculated by the translator from the values given for FeO and Fe.

12 In section 2.1.2 above it is stated that 600-700 kg of charcoal are produced from 2000-2500 kg of wood. These figures suggest a yield of 28-30 per cent rather than the 23 per cent indicated here. Note that the charcoal kiln described in chapter 4 below yields 30-35 per cent.

13 See n. 10.

14 See n. 7.

15 The 'enclosure', *baozi*, is not mentioned elsewhere in the text, and I am not sure what it is. It might be some sort of trough or guide to channel molten iron into the pig mould.

16 The 'utilization coefficient' appears to be an allowance for leakage in the windbox and air-pipes.

17 It is not clear to me how the molten iron could be *seen* dripping down; consider for example figure 8. Note in chapter 5, section 5.5, that this was a difficulty encountered in the study of the Huang Jiguang blast furnace, which is very similar to this one.

18 The figure 4.2% carbon is probably a typographical error for 4.29%. See section 2.3.3 below.

19 Further discussion, on the chemical reactions involved, omitted in the translation.

20 Obviously the text contradicts itself here. The temperature is 'never very high', but 1600° C is very high indeed. This is beyond the melting point of even pure iron, but the iron is 'never completely molten'. '1600' may be a typographical or other error, or it may refer to a maximum reached directly under the tuyère, perhaps very briefly. It is unfortunate that we are not told how the temperature was measured.

21 Further discussion, on the chemical reactions involved, omitted in the translation.
22 A discussion of various recent improvements in the local methods of smelting and refining has been omitted in the translation.

Notes to chapter 3

1 Yang Dajin (1938, pp. 314-5), has some remarks on the decline of the ironsand iron industry in Zhejiang after the introduction of foreign iron.
2 A book titled *Tie ye zhi,* 'Treatise on iron-smelting', was written by Fu Jun, who was in charge of a state ironworks in Zunhua from 1506 to 1521. The book is no longer extant, and we know it only from quotations in other works. These quotations are very confused, but it is at least clear that the ore used was ironsand (Liu Yuncai 1978, p. 25).
3 Yang Dajin's encyclopedic work contains a great deal of information on traditional Chinese industry, but its usefulness is reduced considerably by the fact that it never cites its sources.
4 *Tatara – An old iron-making process of Japan,* 16 mm, colour, 30 min. Tokyo: Iwanami Productions. (Date not given; perhaps in the 1960s.)

Notes to chapter 4

1 Guo Yujing (chapter 5, section 5.4, below) indicates about 3.2 kg. Nyström states that 7.5 kg of charcoal were consumed for each kg of pig-iron; however this figure is likely to be in error, as noted in chapter 5, section 5.3, below.
2 We shall see in chapter 5, section 5.6, below that some of the charcoal leaves the furnace in the slag, and can be returned to the furnace. This could reduce charcoal consumption considerably, perhaps by as much as 50 per cent.
3 *Fujian:* Rosenbluth 1912, p. 655; CEM 1926b, pp. 494, 496, 499; Hurlbut 1930, p. 104. *Gansu:* Rosenbluth 1912, pp. 654-5. *Guangdong:* von Richthofen 1872, pp. 46-7; 1912, p. 444; Shaw 1914, p. 90; Fenzel 1929, pp. 41-3, 45, 78, 91, 92; Pendleton 1937, pp. 474-5, 479. *Hunan:* Rosenbluth 1912, p. 654; Shaw 1914, pp. 59, 61-3. *Shanxi* (?): CEM 1926a, pp. 56, 57. *Sichuan:* C.Y. Hu 1946, p. 106.

Here I have cited only references to traditional forest cultivation practices for the production of timber or fuel. Ornamental trees, mulberry trees for sericulture, bamboo, fruit trees, and industrial species such as the Tung-oil tree have always been cultivated very widely in China.
4 Percy (1875, pp. 384-6) gives a long abstract of Kovanko's article.
5 An article by Kovanko in the same issue of the same journal describes a tour of a coal mine near Beijing.

Notes to chapter 5

1 McCaskey 1903a, pp. 51-5; see also 1903b. Johannsen (1953, p. 35) gives a diagram of this furnace, from McCaskey 1903b; Tylecote (1976, p. 85) gives a simplified version of the same diagram.

2 Several such copper-smelting furnaces, dating from no later than 500 B.C., have been excavated near an ancient copper mine at Tonglüshan, Daye County, Hubei. These are small shaft furnaces, typically 1.5 m high with elliptical cross-section 40×70 cm. See Lu Benshan & Wang Fuguo 1981; Lu Benshan & Hua Jueming 1981; Zhu Yingyao 1981; Bai Rongjin & Yin Weizhang 1982; in English, Xia Nai & Yin Weizhang 1982; Tylecote 1983.

3 The technique of annealing white-cast iron objects to produce malleable cast-iron was developed by about the third century B.C.; with this development cast-iron became a useful material for a very broad class of mass-produced items (see e.g. Li Zhong 1975).

4 Liu Yuncai 1978, p. 23, quoting Chen Yingqi 1959. Other reports of excavations of post-Han ironworks are KG 1965.3: 124-30 and Tang Yunming 1959.

5 Many can be found through citations in Needham 1958, Yang Kuan 1960, Liu Yuncai 1978, and *Yejin jianshi* 1978.

6 The furnace was named for Huang Jiguang (1930-52), a hero of the Korean war.

7 In *Gangtie* (Iron and steel), 1958, no. 15.

8 The rows total 100.15, 99.55, 100.00, and 100.00. At first I thought this indicated that in the first two analyses each component was determined separately; however the second row is repeated later in the article (see below) with the water content given as 2.94% instead of 2.49%. With this correction the row adds up to 100.00%; thus it is likely that 2.49 is a typographical error. If in the first row the carbon content 83.5% is taken to be a typographical error for 83.35% then this row also totals 100.00%.

9 There is something odd about these figures: they imply that the *wet* ironsand was about 64% Fe, whereas the analysis gives only 62.12% for *dry* ironsand. If we assume that the scrap contained 95% iron, which seems more reasonable than 80%, we obtain instead the following inputs to produce 1 kg of new iron:

charcoal	1.3 kg
ironsand	1.7 kg

This implies that the wet ironsand in the furnace charge contained about 58% Fe and about 7% water. These are rough calculations based on uncertain assumptions; they are not to be taken too seriously.

10 *Oxygen in blast:* according to calculations in chapter 2 (section 2.2.5.2), the blast input required to produce 1 kg pig-iron is about 4 kg air. Air is 23% oxygen by weight; this gives 0.9 kg oxygen in the blast. *Oxygen in iron oxides:* according to the analysis in chapter 2 (section 2.2.1.1), the dry ironsand contains 70.24% Fe_3O_4 and 19.55% Fe_2O_3, total Fe 64.55%. The *dry* ironsand thus contains 25.24% oxygen in iron oxides. An input of 2 kg *wet* ironsand is needed to produce 1 kg pig-iron. Assuming a negligible loss of iron in the slag, the wet ironsand thus contains *c.* 50% Fe; comparing this with the Fe content of the dry ironsand indicates 23% water in the wet ironsand. Therefore the wet ironsand contains *c.* 20% oxygen in iron oxides.

11 The possibility of soot in the top-gas, produced by the reaction $2CO \rightarrow CO_2 + C$, is not relevant here, since oxygen is required to produce the CO.

Notes to chapter 6

1 An analysis of this soil (*Lian'gang* 1958, p. 26) is: Al_2O_3 15.12-15.40%, SiO_2 67.78-69.24%, Fe_2O_3 3.03-3.23%, loss on heating 3.46-4.03%. For furnace construction in areas where this type of soil is not obtainable, two substitutes are suggested: (1) a mixture of 45% loess, 36% sand, and 18% fireclay (pp. 26-7); (2) a mixture of 60% yellow clay and 40% coarse sand (p. 11).

2 *Jin* is a traditional unit of weight, sometimes translated 'catty'. Its exact weight has varied considerably from place to place and time to time, but it is now standardized at 0.5 kg. It is not clear whether this standard was followed here. Tegengren (1924) always takes one *jin* to be 0.6 kg, as in the passage quoted further below.

3 Tegengren 1924, pp. 317 (Manchuria), 327-9 (Shanxi), 335-6 (southern Henan), 339 (Hunan), 341 (Guizhou), 346 (Sichuan), 359-64 (Yunnan). See also Yang Kuan 1960, pp. 174-91, and Cremer 1913, pp. 51-2.

4 Rosenqvist 1974, p. 388; cf. Percy 1864, pp. 663, 818. This is of course not true if a basic slag is used; this was Thomas's invention of about 1878.

CHINESE CHARACTERS

In the above all Chinese names and terms have been transcribed according to the Pinyin system. This is the standard transcription for Chinese in Western-language publications in the People's Republic of China, and outside China it is gradually replacing the many other transcriptions now in use. Since Pinyin is not yet very familiar to non-Sinologists, the more common Wade-Giles transcription is also given for each entry in the following list.

Pinyin	*Wade-Giles*	*Chinese characters*
Anhui	An-hui (Anhwei)	安徽
Bai Rongjin	Pai Jung-chin	白荣金
baozi	pao-tzu	包子
Beijing	Pei-ching (Peking)	北京
Cangshan	Ts'ang-shan	苍山
Changjiang	Ch'ang-chiang (Yangtze River)	长江
chao	ch'ao	炒
chi	ch'ih	尺
Chongqing	Ch'ung-ch'ing (Chungking)	重庆
choulu	ch'ou-lu	抽炉
cun	ts'un	寸
Dabieshan	Ta-pieh-shan	大别山
Daye	Ta-yeh	大冶

Pinyin	*Wade-Giles*	*Chinese characters*
E-Yu-Wan	O-Yü-Wan	鄂豫皖
Ėcheng	O-ch'eng	鄂城
Fangcheng	Fang-ch'eng	方城
fanshelu	fan-she-lu	反射炉
fengxiang	feng-hsiang	风箱
Fu Jun	Fu Chün	傅浚
Fujian	Fu-chien (Fukien)	福建
Funiushan	Fu-niu-shan	伏牛山
Gang tie shengchan da yuejin lunwen xuanji	Kang t'ieh sheng-ch'an ta yüeh-chin lun-wen hsüan-chi	钢铁生产大跃进论文选集
Gangtie	Kang-t'ieh	钢铁
Gansu	Kan-su	甘肃
Gongxian	Kung-hsien	巩县

Pinyin	*Wade-Giles*	*Chinese characters*
Gongxian Tieshenggou	Kung-hsien T'ieh-sheng-kou	巩县铁生沟
Guangdong	Kuang-tung (Kwangtung)	广东
Guangming ribao	Kuang-ming jih-pao	光明日报
Guangshan	Kuang-shan	光山
Guanyin mian	Kuan-yin mien	观音面
Guizhou	Kuei-chou (Kweichow)	贵州
guo	kuo (wok)	锅
Guo Yujing	Kuo Yü-ching	郭玉璟
Guomindang	Kuo-min-tang	国民党
Guxingzhen	Ku-hsing-chen	古荥镇
Han	Han	汉
Hankou	Han-k'ou (Hankow)	汉口
Hanyang	Han-yang	汉阳
Hebei	Ho-pei (Hopeh)	河北

Pinyin	*Wade-Giles*	*Chinese characters*
Hefei	Ho-fei	合肥
Henan	Ho-nan	河南
Henan sheng dizhi diaochasuo huikan	Ho-nan sheng ti-chih tiao-ch'a-so hui-k'an	河南省地質調查所彙刊
Hongdaoyuan	Hung-tao-yüan	宏道院
Hu Zhaoliang	Hu Chao-liang	胡兆量
Hua Jueming	Hua Chüeh-ming	华觉明
Huang Jiguang	Huang Chi-kuang	黄继光
Huang Zhanyue	Huang Chan-yüeh	黄展岳
Hubei	Hu-pei	湖北
Hunan	Hu-nan	湖南
Jiangxi	Chiang-hsi (Kiangsi)	江西
Jiayu	Chia-yü	嘉鱼

Pinyin	*Wade-Giles*	*Chinese characters*
jieshu	chieh-shu	檞树
jin	chin	斤
Kaogu	K'ao-ku	考古
Kaogu xuebao	K'ao-ku hsüeh-pao	考古学报
Kuangye	K'uang-yeh	鑛冶
Kuangye banyuekan	K'uang-yeh pan-yüeh-k'an	鑛冶半月刊
Li Longqian	Li Lung-ch'ien	李龙潜
Li Zhong	Li Chung	李众
Lijiazhai	Li-chia-chai	李家寨
Linyi	Lin-i	临沂
lishu	li-shu	栎树
Liu Yuncai	Liu Yün-ts'ai	刘云彩
Liu Zhichao	Liu Chih-ch'ao	刘志超
Lu Benshan	Lu Pen-shan	卢本珊

Pinyin	*Wade-Giles*	*Chinese characters*
Ma Hong	Ma Hung	马洪
Macheng	Ma-ch'eng	麻城
malishu	ma-li-shu	麻栗树
Mancheng	Man-ch'eng	满城
Mancheng Han mu fajue baogao	Man-ch'eng Han mu fa-chüeh pao-kao	满城汉墓 发掘报告
Mao Zedong	Mao Tse-tung	毛泽东
Nanjing	Nan-ching (Nanking)	南京
Nanyang	Nan-yang	南阳
Qishui	Ch'i-shui	蕲水
Quanguo xin shu mu	Ch'üan-kuo hsin shu mu	全国新书目
Quanzhou	Ch'üan-chou	泉州
Shaanxi (Shănxi)	Shan-hsi (Shensi)	陕西

Pinyin	*Wade-Giles*	*Chinese characters*
Shandong	Shan-tung	山东
Shangcheng	Shang-ch'eng	商城
Shanxi (Shānxi)	Shan-hsi (Shansi)	山西
Shuangshi	Shuang-shih	双狮
Sichuan	Ssu-ch'uan (Szechwan)	四川
Song	Sung	宋
Song Yingxing	Sung Ying-hsing	宋應星
Sun Shangqing	Sun Shang-ch'ing	孙尚清
tanfen	t'an-fen	炭粉
Tang	T'ang	唐
Tang Youyu	T'ang Yu-yü	唐有余
Tang Yunming	T'ang Yün-ming	唐云明
tatara (Japanese word)		踏鞴
Tengxian	T'eng-hsien	滕县

Pinyin	*Wade-Giles*	*Chinese characters*
Tian gong kai wu	T'ien kung k'ai wu	天工開物
Tie ye zhi	T'ieh yeh chih	鐵冶志
Tieshenggou	T'ieh-sheng-kou	铁生沟
Tongbaishan	T'ung-pai-shan	桐柏山
Tonglüshan	T'ung-lü-shan	铜绿山
Tufa diwen lian'gang	T'u-fa ti-wen lien-kang	土法低温炼钢
Tufa liantie	T'u-fa lien-t'ieh	土法炼铁
Wafangzhuang	Wa-fang-chuang	瓦房庄
Wang Fuguo	Wang Fu-kuo	王富国
Wang Mang	Wang Mang	王莽
Wang Zhenduo	Wang Chen-to	王振铎
Wenwu	Wen-wu	文物
Wuwei	Wu-wei	武威
Xi'an	Hsi-an (Sian)	西安

Pinyin	*Wade-Giles*	*Chinese characters*
Xia Nai	Hsia Nai	夏鼐
Xia Xiangrong	Hsia Hsiang-jung	夏湘蓉
Xiandai Zhongguo shiye zhi	Hsien-tai Chung-kuo shih-yeh chih	現代中國實業誌
Xinyang	Hsin-yang	信阳
Xishui	Hsi-shui	浠水
Yangcheng	Yang-ch'eng	阳城
Yang Dajin	Yang Ta-chin	楊大金
Yang Kuan	Yang K'uan	杨宽
Ye Jun	Yeh Chün	冶军
Yejin bao	Yeh-chin pao	冶金报
Yichang	I-ch'ang	宜昌
Yin Weizhang	Yin Wei-chang	殷玮璋
Youse jinshu	Yu-se chin-shu	有色金属

Pinyin	*Wade-Giles*	*Chinese characters*
yuan	yüan	元
Yunnan	Yün-nan	云南
Zhang Youxian	Chang Yu-hsien	張友賢
Zhang Zongling	Chang Tsung-ling	张宗龄
Zhaohecun	Chao-ho-ts'un	赵河村
Zhejiang	Che-chiang (Chekiang)	浙江
Zhengzhou	Cheng-chou	郑州
Zhong Guangyan	Chung Kuang-yen	钟广言
Zhongguo gudai kuangye kaifa shi	Chung-kuo ku-tai k'uang-yeh k'ai-fa shih	中国古代矿业开发史
Zhongguo jingji jiegou wenti yanjiu	Chung-kuo ching-chi chieh-kou wen-t'i yen-chiu	中国经济结构问题研究

Pinyin	*Wade-Giles*	*Chinese characters*
Zhongguo shehui jingji shi luncong	Chung-kuo she-hui ching-chi shih lun-ts'ung	中国社会经济史论丛
Zhongguo tufa yetie lian'gang jishu fazhan jianshi	Chung-kuo t'u-fa yeh-t'ieh lien-kang chi-shu fa-chan chien-shih	中国土法冶铁炼钢技术发展简史
Zhongguo yejin jianshi	Chung-kuo yeh-chin chien-shih	中国冶金简史
Zhonghua Minguo wu-nian di-wu-ci nong shang tongji biao	Chung-hua Min-kuo wu-nien ti-wu-tz'u nung shang t'ung-chi piao	中華民國五年第五次農商統計表
Zhu Yingyao	Chu Ying-yao	朱英尧
Zunhua	Tsun-hua	遵化

REFERENCES AND ABBREVIATIONS

Alley, Rewi 1961a *China's hinterland—in the leap forward.* Peking: New World Press.

—— 1961b *Together they learnt how to make iron and steel. Some early types of furnaces used in 1958-9, in China.* A book of pictures by —. (An unpublished album of 299 photographs; this and the original negatives are in the collection of the East Asian History of Science Library, Cambridge).

Bai Rongjin & Yin Weizhang 1982 'Hubei Tonglüshan gu tongkuang zaici fajue—Dong Zhou liantonglu de fajue he liantong moni shiyan' (Further investigation of the ancient copper mine at Tonglüshan, Daye County, Hubei: Excavation of an ancient copper-smelting furnace and copper-smelting simulation experiments), KG 1982.1: 18-22 + plate 5.

Buck, David D., tr. 1975 'Reconnaissance of ancient mine and smelter sites in Hupei province', *Chinese sociology and anthropology, 8.*1: 3-18. Translation of KG 1974.4: 251-4, 256.

CEM = *Chinese economic monthly.*

CEM 1926a 'Survey of forestry conditions and lumber production in Shansi', *3.*2: 55-60.

CEM 1926b 'Timber trade in Fukien', *3.*11: 494-500.

Chambers, [William] 1757 *Desseins des édifices, meubles, habits, machines, et ustenciles des chinois.* Gravés sur les originaux dessinés à la Chine par Mr. Chambers, architecte ... Auxquels est ajoutée une description de leurs temples, de leurs maisons, de leurs jardins, &c. Londres: J. Haberkorn. (There is also an English-language edition, which I have not seen.)

Chen, Chi-yün 1939 'Über die forstlichen Verhältnisse in der mittelchinesischen Provinz Anhui', *Zeitschrift für Weltforstwirtschaft* 7.2: 55-66.

Chen Yingqi 1959 'Handan kuangshancun faxian Song dai yetielu' (An iron-smelting furnace of the Song period discovered at Kuangshancun, Handan, Hebei), GMRB 1959.12.13: 3. [Not seen.]

Cheng Shih-po 1978 'An iron and steel works of 2,000 years ago', *China reconstructs 27.*1: 32-34.

CN = *Chemical news.*

CN 1884 'Chinese iron foundries and rice pan casting', by T.I.B., *50.*1287: 40-41 (July 25, 1884). Abstract in *Engineering and mining journal, 39.*16: 264 (April 18, 1885).

Corpus 1950-51 *Corpus des pierres sculptés Han (estampages),* 1-2. Pékin: Centre d'Études Sinologiques. Loose plates + French & Chinese text in two cases. Chinese title: *Han dai huaxiang quanji.*

Cremer, L. 1913 'Bericht über eine Reise in der chinesischen Provinz Szetschuan', *Zeitschrift für das Berg-, Hütten- und Salinenwesen im preussischen Staate, 61:* 1-146.

Ewbank, Thomas 1849 A descriptive account of hydraulic and other machines ..., 3rd ed. New York. [1st ed. 1842.]

Fenzel, G. 1929 'On the natural conditions affecting the introduction of forestry as a branch of rural economy in the province of Kwangtung, especially in North Kwangtung', *Lingnan science journal 7:* 37-97 (June 1929).

Ganze Volk 1959 *Das ganze Volk schmilzt Eisen und Stahl.* Peking. Unpaginated. (The English edition, *All China makes iron and steel,* Peking 1958, has not been available to me.)

GMRB = *Guangming ribao* (Guangming daily).

Gongxian 1962 *Gongxian Tieshenggou* (Excavation of a Han-period ironworks site at Tieshenggou, Gongxian County, Henan). Beijing. 40 pp. + 16 plates. English abstract pp. 39-40. English title: *T'ieh Sheng Kou, Kung Hsien.*

Gordon, Robert B. 1983 'Materials for manufacturing: the response of the Connecticut iron industry to technological change and limited resources', *Technology and culture, 24.*4: 602-34.

Guo Yujing 1932 'Henan kuangchan' (Mineral production of Henan province), *Henan sheng dizhi diaochasuo huikan* (Memoirs of the Geological Survey of Henan Province), *1:* 159-220.

Hartwell, Robert 1962 'A revolution in the iron and coal industries during the Northern Sung', *Journal of Asian studies 21.*2: 153-162.

—— 1966 'Markets, technology, and the structure of enterprise in the development of the eleventh-century Chinese iron and steel industry', *Journal of economic history 26.*1: 29-58.

—— 1967 'A cycle of economic change in Imperial China: coal and iron in northeast China, 750-1350', *Journal of the economic and social history of the Orient, 10:* 102-59.

Hayes, James 1971 'Charcoal burning in Hong Kong', *Journal of the Hong Kong Branch of the Royal Asiatic Society, 11:* 199-203.

Ho, Ping-ti 1959 *Studies on the population of China, 1368-1953.* Cambridge, Mass.: Harvard University Press.

Hommel, Rudolph P. 1937 *China at work. An illustrated record of the primitive industries of China's masses, whose life is toil, and thus an account of Chinese civilization.* New York. (Repr. Cambridge, Mass.: MIT Press.)

Hsia Nai 1975 'The slaves were the makers of history', *China reconstructs, 24.*11: 40-43.

Hsieh, C.Y. 1935 'Classification of iron deposits in China', appendix, pp. 163-91 in his *Geology of the iron deposits in the lower Yangtze region* (*Memoirs of the Geological Survey of China,* Series A, no. 13). Peiping.

Hu, Charles Y. 1946 *The agricultural and forestry land-use of Szechuan Basin.* Chicago.

Hu, H.H. 1929 'The nature of the forest flora of southeastern China', *Bulletin of the Peking Society of Natural History, 4.*1: 47-56 (Sept. 1929).

Hu Zhaoliang 1981 'Gangtie gongye jiegou' (The structure of the iron and steel industry), pp. 295-316 in Ma Hong & Sun Shangqing, eds., *Zhongguo jingji jiegou wenti yanjiu* (Problems of China's economic structure), 2 vols., Beijing.

Huang, C.T. 1919 'Conversion of white iron into foundry: How Chinese native irons may be made available as a means of relieving the scarcity of other grades in that country', *Iron age* (New York), *103.*4: 231-2 (Jan. 23, 1919).

Huang Zhanyue 1976 'Guanyu Zhongguo kaishi yetie he shiyong tieqi de wenti' (On the problem of the first smelting of iron and the first use of iron implements in China), WW 1976.8: 62-70.

Hurlbut, Floy 1930 *The Fukienese: A study in human geography*. Ph.D. thesis, University of Nebraska.

Inouye, Kinosuke 1912 'On some iron ores in China', *Compte rendu de la XI:e session du Congrès Géologique International, 1910.* Stockholm.

Irresberger, C. 1916 'Chinesische Kochpfannen mit ausserordentlich geringer Wandstärke', *Stahl und Eisen 36:* 319.

Johannsen, Otto 1916 [Comments on Irresberger 1916], *Stahl und Eisen 36:* 417.

—— 1953 *Geschichte des Eisens.* Düsseldorf.

KG = *Kaogu* (Archaeology)

KG 1965.3: 124-30 + plates 6-7 'Heilongjiang Acheng xian Xiaolin diqu Jin dai yetie yizhi' (A Jin-period [A.D. 1115-1234] ironworks site in Xiaoling Prefecture, Acheng County, Heilongjiang), by the Heilongjiang Provincial Museum.

KG 1978.6: 400-1 + plate 10 'Cixian Yuan dai muchuan chutu tieqi jinxiang jianding' (Metallographic examination of iron artifacts from the Yuan-period shipwrecks found in Cixian County, Hebei).

KGXB = *Kaogu xuebao* ('Chinese journal of archaeology').

KGXB 1978.1: 1-24 + plates 1-2 'Henan Han dai yetie jishu' (The iron and steel making techniques of the Han dynasty in Honan).

Kovanko, M. 1838 'De la fabrication du charbon de bois, en Chine', *Annuaire du Journal des mines de Russie,* Année 1838 (Saint-Pétersbourg 1840), pp. 375-80 + pl. 2.

Lau, D.C., tr. 1970 *Mencius.* Penguin Books.

Li Jinghua 1974 'Han dai tie nongqi mingwen shishi' (Explication of the inscriptions on Han-period iron agricultural implements), KG 1974.1:61-66, 60.

Li Longqian 1981 'Qing dai qianqi Guangdong caikuang yezhu ye zhong de zibenzhuyi mengya' (Embryonic capitalism in the mining and metallurgical industry of Guangdong in the early Qing period), in *Shehui jingji* 1981, pp. 352-80.

Li Zhong 1975 'Zhongguo fengjian shehui qianqi gangtie yelian jishu fazhan de tantao' (The development of iron and steel technology in ancient China), KGXB 1975.2: 1-20, English summary 21-2, plates 1-6.

Lian'gang 1958 *Tufa diwen lian'gang* (Local methods of low-temperature steel-refining), comp. by Jiangsu Provincial Office of Metallurgical Industry, 50 pp., publ. in Nanjing, Oct. 1958, in an edition of 5000 copies.

Lin, D.Y. 1943 'Forestry', pp. 515-29 in *The Chinese year book,* 6th issue, prepared by The Council of International Affairs, Chungking. Publ. in Bombay and London.

Little, H.A. 1908 'Report on the physical, commercial, social and general conditions of Ichang and neighbourhood', *Diplomatic and consular reports, Miscellaneous series,* no. 671. Ed. at the Foreign Office and the Board of Trade. London.

Liu Xinjian & Chen Zijing 1974 'Shandong Cangshan faxian Dong Han Yongchu jinian tiedao' (An iron knife dated AD 112 discovered in Cangshan County, Shandong), WW 1974.12: 61 + plate 5.

Liu Yuncai 1978 'Zhongguo gudai gaolu de qiyuan he yanbian' (The origin and development of the blast furnace in ancient China), WW 1978.2: 18-27.

Liu Zhichao & Tang Youyu 1959 'Hubei sheng Macheng xian Huang Jiguang lu gaochan jingyan' (High production with the Huang Jiguang blast furnace in Macheng County, Hubei), *Gangtie* (Iron and steel), 1959.6: 182-7.

Lu Benshan & Hua Jueming 1981 'Tonglüshan Chunqiu liantong shulu de fuyuan yanjiu' (Experimental reconstruction of a copper-smelting shaft furnace of the Spring and Autumn period at Tonglüshan, Daye County, Hubei), WW 1981.8: 40-45.

Lu Benshan & Wang Fuguo 1981 'Hubei Tonglüshan Chunqiu shiqi liantong yizhi fajue jianbao' (Excavation of a copper-smelting site of the Spring and Autumn period at Tonglüshan, Daye County, Hubei), WW 1981.8: 30-39.

Lux, Fr. 1912 'Koksherstellung und Hochofenbetrieb im Innern Chinas', *Stahl und Eisen 32:* 1404-7.

McCaskey, H.D. 1903a 'Report on a geological reconoissance of the iron region of Angat, Bulacan', *The Mining Bureau* [Philippine Islands]. *Bulletin* no. 3. Manila: Bureau of Public Printing.

—— 1903b 'Iron metallurgy in the Philippines', *Engineering and mining journal,* 21 Nov. 1903, p. 780.

MacFarquhar, Roderick 1983 *The origins of the Cultural Revolution.* Vol. 2: *The Great Leap Forward.* Oxford University Press.

Mancheng 1980 *Mancheng Han mu fajue baogao* (Excavation of the Han tombs at Mancheng), 2 vols., Beijing. English abstract pp. 449-51.

Mell, R. 1933 'Die ehemalige Waldverbreitung in China auf Grund der Verbreitung von Waldtieren', *Zeitschrift der Gesellschaft für Erdkunde zu Berlin,* Heft 3/4: 101-8.

Mott, R.A. 1983 *Henry Cort: The great finer. Creator of puddled iron.* Ed. by Peter Singer. London: The Metals Society.

NED = *Notes et études documentaires* (La documentation française).

NED 1959 'L'industrie sidérurgique chinoise (1890-1959)', no. 2591, 12 novembre 1959.

Needham, Joseph 1958 *The development of iron and steel technology in China* (Second Biennial Dickinson Memorial Lecture to the Newcomen Society 1956). London.

—— 1965 *Science and civilisation in China,* vol. 4, part 2. Cambridge.

Needham, Joseph & Dorothy, eds. 1948 *Science outpost: Papers of the Sino-British Science Co-Operation Office, 1942-1946.* London.

Nikolayev, S.A. & Molodtsova, L.I. 1960 'The present state of the Chinese iron and steel industry', *Soviet geography: Review and translation* (New York: American Geographical Society), *1*.8: 55-71. (Translated from *Geografiya i khozyaystvo,* 1960, no. 6, pp. 34-44.)

Nong shang 1919 *Zhonghua Minguo wu-nian di-wu-ci nong shang tongji biao* (Statistics of agriculture and commerce, fifth report, 1916), comp. by the Chinese Ministry of Agriculture and Commerce. n.p.

Nyström, Erik 1953 'Naturtilgångar i Kina', *Clarté* (Stockholm), *26*.4/5: 25-7.

—— 1956 'Malmgeologiska expeditionen til Kina 1914', *Daedalus: Tekniska Museets årsbok* (Stockholm), pp. 142-7.

den Ouden, Alex 1981-2 'The production of wrought iron in finery hearths', parts 1-2. *Historical metallurgy: Journal of the Historical Metallurgy Society, 15*.2: 63-87; *16*.1: 29-32.

Pendleton, Robert L. 1937 'Forestry in Kwangtung Province, South China', *Lingnan science journal 16*.3: 473-80 + plates 8-15.

Percy, John 1864 *Metallurgy: The art of extracting metals from their ores, and adapting them to various purposes of manufacture.* [Vol. 2:] *Iron and steel.* London.

—— 1875 *Metallurgy: The art of extracting metals from their ores.* [Vol. 1:] *Introduction, refractory materials, and fuel.* Rev. and greatly enlarged ed., London. [1st ed. 1861].

Perkins, Dwight, ed. 1977 *Rural small-scale industry in the People's Republic of China.* Berkeley. (Paperback ed. 1981.)

Resolution 1981 *Resolution on CPC history* [Communist Party of China]. Beijing.

Richardson, S.D. 1966 *Forestry in Communist China.* Baltimore.

von Richthofen, Ferdinand [*c.* 1872] *Baron Richthofen's letters, 1870-1872* [to the Shanghai Chamber of Commerce]. Shanghai, n.d.

—— 1912 *China: Ergebnisse eigener Reisen und darauf gegründeter Studien.* Bd. 3: *Das südliche China.* Berlin.

Rosenbluth, R. 1912 'Forests and timber trade of the Chinese Empire', *Forestry quarterly 10*.4: 642-72.

Rosenqvist, Terkel 1974 *Principles of extractive metallurgy.* New York.

Shaw, Norman 1914 *Chinese forest trees and timber supply.* London & Leipsic.

Shehui jingji 1981 *Zhongguo shehui jingji shi luncong* (Essays on Chinese social and economic history), no. 1. Taiyuan: Shanxi Renmin Chubanshe.

Shiba Yoshinobu 1970 *Commerce and society in Sung China,* tr. by Mark Elvin. (*Michigan abstracts of Chinese and Japanese works on Chinese history,* no. 2). Ann Arbor, Mich.

Sun, E-tu Zen & Sun, Shiou-chuan (tr.) 1966 *T'ien-kung k'ai-wu: Chinese technology in the seventeenth century,* by Sung Ying-hsing. University Park, Pa. [Translation of TGKW.]

Tang Yunming 1959 'Hebei Xingtai faxian Song mu he yetie yizhi' (A tomb and an iron-smelting site of the Song period [A.D. 960-1279] discovered in Xingtai County, Hebei), KG 1959.7: 369.

Tegengren F.R. 1924 *The iron ores and iron industry of China. Including a summary of the iron situation of the circum-Pacific region.* Part I, 1921-23; part II, 1923-24. (*Memoirs of the Geological Survey of China,* series A, no. 2). English text and Chinese summary, each with continuous pagination through the two volumes. Photographic reprint available from University Microfilms, Ann Arbor, Michigan; order no. OP 12715.

TGKW = *Tian gong kai wu* (The products of Heaven and man, by Song Yingxing, 1st ed. dated 1637). Typeset edition ed. by Zhong Guangyan, Beijing; repr. Hong Kong 1978. (Translation Sun & Sun 1966).

Tufa 1958 *Tufa liantie* (Iron production by local techniques), ed. by Henan Provincial Office of Metallurgy, 26 pp., publ. in Zhengzhou, Henan, August 1958, in an edition of 100 085 copies. (Translated in chapter 2 above).

Tylecote, R.F. 1976 *A history of metallurgy.* London.

—— 1983 'Ancient metallurgy in China', *The metallurgist and materials technologist,* Sept. 1983, pp. 435-9.

Wales, Nym [pseud. of Helen Foster Snow] 1952 *Autobiographies of Chinese Communists.* Stanford University Press.

Wang Zhenduo 1959 'Han dai yetie gufengji de fuyuan' (Reconstruction of the iron-smelting bellows of the Han period), WW 1959.5: 43-4.

Wen Huanran & He Yeheng 1982 'China's forests—past and present', *China reconstructs,* 1982.2: 65-6.

WW = *Wenwu* (Cultural relics)

WW 1959.1: 2 'Han huaxiang yetie tu shuoming' (A Han relief showing iron production), by Shandong Provincial Museum.

WW 1975.2: 1-12 'Hubei Tonglüshan Chunqiu Zhanguo gu kuangjing yizhi fajue jianbao' (Excavation of the mining and foundry sites of the Spring and Autumn period and the Warring States Period at T'ung-lu-shan in Hupei province).

WW 1978.10:44-7 'Changsha xin faxian Chunqiu wanqi de gang jian he tieqi' (A steel sword and iron vessels of the late Spring and Autumn period newly discovered in Changsha), by the Archaeological Team for the Construction of the Changsha Railway Station.

Xia Nai & Yin Weizhang 1982 'Digging up an ancient copper mine', *China reconstructs, 31*.3: 38-40 (March 1982).

Xia Xiangrong 1980 *Zhongguo gudai kuangye kaifa shi* (The history of the development of the mining industry in ancient China), by —, Li Zhongjun, and Wang Genyuan. Beijing.

Yang Dajin 1938 *Xiandai Zhongguo shiye zhi* (Compendium on modern Chinese commerce), vol. 2. Changsha.

Yang Kuan 1960 *Zhongguo tufa yetie lian'gang jishu fazhan jianshi* (A short history of the development of Chinese local siderurgical techniques). Shanghai.

—— 1982 *Zhongguo gudai yetie jishu fazhan shi* (The development of siderurgical technology in ancient China). Shanghai: Renmin Chubanshe.

Ye Jun 1975 'Tonglüshan gu kuangjing yizhi chutu tiezhi ji tongzhi gongju de chubu jianding' (Preliminary analysis of some bronze and iron implements from the ancient mine site at Tonglüshan, Daye county, Hubei), WW 1975.2: 19-25.

Yejin jianshi 1978 *Zhongguo yejin jianshi* (A concise history of metallurgy in China). Beijing.

Yuejin 1958 *Gang tie shengchan da yuejin lunwen xuanji* (Collected essays on the Great Leap in iron and steel production). Beijing: Kexue Chubanshe.

Zhang Youxian & Guo Yujing 1932 'Henan tiekuang' (Iron ores of Henan), *Henan sheng dizhi diaochasuo huikan* (Memoirs of the Geological Survey of Henan Province), *1:* 221-43.

Zhang Zongling 1957 'Dui jianshe zhong xiao xing gangtiechang guanche qinjian jieyue fangzhen de jidian yijian' (Thoughts on the implementation of the policy of 'industriousness and thrift' in the establishment of small- and medium-scale ironworks), *Yejin bao* (Metallurgical news), 1957.22: 16-17.

Zhu Xingzhong 1932 'Hebei ge kuang gaiyao' (Mines and mining in Hopei province), part 1, *Kuangye* (Mining and metallurgy), *5*.17: 1-9, 1-74 (two paginations) + foldout map.

Zhu Yingyao 1981 'Tonglushan gu kuangye yizhi liantong jishu de tantao' (Discussion on copper smelting technique at the ancient Tonglushan mine smelting ruins), *Youse jinshu* (Nonferrous metals), *33*.4: 63-7. Very brief English abstract, p. 67.

INDEX